Notes for New Beekeepers

Bill Cadmore

Northern Bee Books

Notes for New Beekeepers

© Bill Cadmore

All rights reserved. No part of this publication may be reproduced, stored in a retrieval system, transmitted in any form or by any means electronic, mechanical, including photocopying, recording or otherwise without prior consent of the copyright holders.

ISBN 978-1-914934-45-2

Published by Northern Bee Books, 2022
Scout Bottom Farm
Mytholmroyd
Hebden Bridge
HX7 5JS (UK)

Design and artwork by DM Design and Print

Notes for New Beekeepers

Bill Cadmore

Content

The Hymenoptera	Page 1
Apis mellifera	Page 3
Three Castes of Honeybee	Page 5
The Life Cycle of the Honeybee	Page 9
Beehives	Page 16
Polystyrene or Wooden Hives?	Page 22
The Parts of a Beehive	Page 24
Bee Space	Page 28
Frames, Combs and Foundation	Page 29
Smokers, Tools and Silly Hats	Page 31
Your Apiary	Page 34
Three Miles	Page 36
Selecting an Apiary Site	Page 37
The Apiary Environment	Page 39
Bees and Neighbours	Page 42
Starting Beekeeping	Page 45
Purchasing Bees	Page 48
Transfer Bees into Your Hive	Page 50
Inspecting the Colony	Page 52
Supering	Page 56
Operate a Two Hive System	Page 57
Queen Cells and Queen Cups	Page 59
Swarm Prevention	Page 62
What to do if you find a queen cell	Page 64
Swarm Control	Page 65
Finding a sealed queen cell - Your bees have swarmed	Page 73
Collecting a Swarm	Page 74

Bee Diseases & Pests	Page 78
The Diseases and Pests that cause problems	Page 80
Brood Diseases	Page 81
Varroasis	Page 82
Adult Diseases	Page 84
Management Issues	Page 85
Pests that cause problems for honeybees	Page 86
Foraging	Page 88
Bee Plants and Flowers	Page 91
Feeding Honeybees	Page 93
Hive Products	Page 96
Honey	Page 99
Extracting and Bottling Techniques	Page 101
Selling Honey	Page 103
The Beekeeping Year	Page 104
New Beekeepers Checklist	Page 109

Useful web links

National Bee Unit Register your apiary here Bee Inspectors give free advice Useful leaflets	https://secure.fera.defra.gov.uk/beebase/index.cfm?sectionid=43
The British Beekeepers Association	http://www.britishbee.org.uk/
The Scottish Beekeepers Association	http://www.scottishbeekeepers.org.uk/
The Welsh Beekeepers Association	http://www.wbka.com/
Royal Entermological Society	http://www.royensoc.co.uk/
Information on Bumblebees	https://www.bumblebeeconservation.org

The Hymenoptera - Bees, Wasps and Ants

Honeybees are insects:- They have an exoskeleton, a body consisting of 3 parts, Head, Thorax and Abdomen. They have 3 pairs of jointed legs. They are closely related to all the other types of bees, the wasps and the ants, all of which are grouped as the Hymenoptera.

Hymenoptera characteristically have two pairs of wings, a large fore pair and a smaller hind pair. These wings are held together by a series of hooks (called hamuli) and may appear like a single pair to the naked eye. Hymenoptera also tend to have prominent antennae, generally with nine or more segments and biting mouthparts.

There are more than 200 species of bees in the British Isles.

Most are solitary bees - They construct their nests by tunnelling in the earth or by utilizing pre-existing cavities in rotten wood, plant stems or beetle burrows in timber. Some mining bees are social but most are not.

Some are Bumblebees – The queen establishes a nest in spring and the colony develops and expands as the summer progresses. New queens are raised and mated before winter. They are social. Only the queen bees survive through winter.

Just one species in the UK is a honeybee and only nine worldwide. They live in colonies with large populations throughout the year. This makes them special because they need to store honey to provide food for the colony during the winter.

In all types of bee the adults and larvae eat nectar, honey and ingest pollen. The pollen is usually collected by females in the branched hairs of the body while the nectar is collected in the 'honey stomach'.

2 Notes for new beekeepers

Classification

Kingdom	Animalia	(Animals)
Phylum	Arthropoda	(Arthropods)
Subphylum	Hexapoda	(Hexapods)
Class	Insecta	(Insects)
Order	Hymenoptera	(Ants, Bees, Wasps and Sawflies)
Family	Ants	
Family	Wasps	
Family	Apidae (Bees)	
Genus Apis	**(Honeybees)**	

    Subgenus   *Micrapis* (Dwarf Honeybees)
                           Apis andreniformis
                           Apis florea
    Subgenus   *Megapis* (Giant Honeybees)
                           Apis dorsata (Asian Honeybee)
    Subgenus   Apis:
                           Apis cerana   (Asiatic Honeybee)
                           Apis koschevnikovi   (Indonesian honeybee)
                           Apis nigrocincta   (Philippine Honeybee)
                           Apis mellifera   **(Western European Honeybee)**

Apis mellifera
(Western European Honeybee)

Apis mellifera is native to Europe, western Asia and Africa. Human introduction of *Apis mellifera* to other continents began in the 17th century. They are now found all around the world, including East Asia, Australia and North and South America.

European honeybees prefer habitats that have an abundant supply of suitable flowering plants, such as meadows, open wooded areas, and gardens. They can survive in grasslands, deserts and wetlands if there is sufficient water, food, and shelter. In the wild they need cavities (e.g. in hollow trees) to nest in.

Apis mellifera are fairly dull in colouration ranging from black, through chestnut to greyish. They have striped abdomens with black/orange/yellow/grey rings. They have hair on their thorax and less hair on their abdomen. They also have a pollen basket on their hind legs. Honeybee legs are mostly dark brown/black.

There are two castes of females. Sterile workers are approx. 10-15 mm long while the fertile queens are larger at 18-20 mm. Males, called drones, are 15-17 mm long at maturity. Though smaller, workers have longer wings than drones. Both castes of females have a stinger that is formed from modified ovipositor structures. In workers the sting is barbed and tears away from the body when used. In both castes the stinger is supplied with venom from glands in the abdomen.

Males have much larger eyes than females, probably to help locate flying queens during mating flights. Males do not have a stinger.

There are several subspecies of *Apis mellifera*. The subspecies vary in their abilities to prosper in different environmental conditions. Some subspecies have the ability to tolerate warmer or colder climates. Subspecies may also vary in their defensive behaviour, tongue length, wingspan and coloration. Abdominal banding patterns also differ - some darker and some with more of a mix between darker and lighter banding patterns. As a general rule subspecies living in colder climates store more honey and swarm less often.

4 Notes for new beekeepers

In the UK the most common subspecies are

Apis mellifera ligustica - The Italian bee.
They are gentle, not very likely to swarm, and produce a large surplus of honey (So many people tell me – My experience is that they produce little honey). They have few undesirable characteristics. Colonies tend to maintain larger populations through winter so they require more winter stores (or feeding) than other temperate zone subspecies. The Italian bee is light coloured with a yellow/golden hue.

Apis mellifera carnica - Carniola region of Slovenia, the Eastern Alps, and northern Balkans.
Better known as the Carniolan Honeybee - popular with beekeepers due to its extreme gentleness. The Carniolan tends to be quite dark in colour. The colonies are known to shrink to small populations over winter and build up very quickly in spring. It is a mountain bee in its native range and is a good bee for colder climates. I have found that they are fairly swarmy.

Apis mellifera caucasia - Caucasus Mountains
This sub-species is regarded as being very gentle and fairly industrious. Some strains are excessive propolisers. It is a large honeybee of sometimes grayish colour.

Apis mellifera mellifera - the dark bee of northern Europe or English Black Bee.
These small, dark-coloured bees. They are good honey producers but can be a little 'testy'.

MY HONEYBEES – Most of us keep mongrel hybrids of these subspecies suited to the area we live in.

Three Castes in the Honeybee Colony

A honeybee is an individual that exists as a part of the cohesive unit called the colony – the superorganism. In the colony each individual works cooperatively for the benefit of all. Within the colony there are three castes of honeybee; The Queen, Workers and Drones.

Workers around the queen

6　Notes for new beekeepers

A Drone

The Work of the three castes within the colony

Workers take on age related tasks after they emerge

1 - 3 days	Cell cleaning and brood incubation
4 - 6 days	Feeding older larvae (honey and pollen)
7 - 12 days	Feeding young larvae (brood food)
13 - 18 days	Processing nectar into honey, wax making, water evaporation, pollen packing.
19 - 21 days	Guarding and starting to forage
22nd day +	Foraging for nectar, pollen, water and propolis.

These times are approximate, older bees can revert to their former tasks should the need arise. Other tasks include housekeeping, ventilation, humidity and temperature control.

Drones are produced in spring and are essential for colony cohesiveness in summer

1 - 11 days	Mainly confined to the hive, only leaving for cleansing flight
12 - 14 days	Sexually Mature - Ready to mate
Autumn	Evicted by the workers - Die

8 Notes for new beekeepers

Queens are produced when the colony is ready to reproduce by swarming

1 - 2 days	Seek out rivals and eliminates them – or form a secondary swarm
3 - 5 days	Orientation flights to locate the hive.
6 - 21 days	Multiple mating flights with many Drone partners.
	3 - 4 days later she will commence egg laying
22nd day +	She remains in the hive for egg laying and pheromone production.
	In 2nd or 3rd year might leave the hive in prime swarm

At the height of the season the colony will contain:
one queen
500 drones
60,000 workers.

Of these workers only a small percentage will actually be foraging at any one time. Most workers will be busy within the hive.

The Life Cycle of The Honeybee

7. Young Bee emerges

2. Egg Hatches

3. Early Larva

4. Fully Grown larva
5. Capped Larval

In cocoon

6. Metamorphosis Completed

1. Queen Lays egg

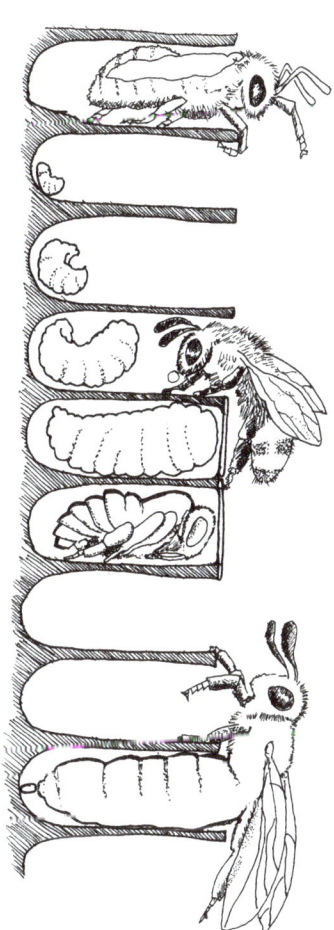

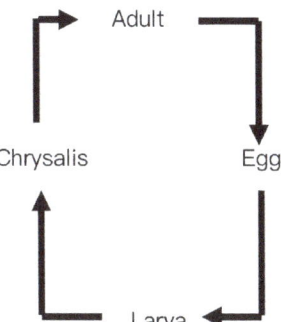

As fully metamorphic insects bees go through four stages in their life cycle - Egg - Larva - Pupa - Adult. The Queen lays eggs in the comb. The time taken for each stage is given below:

	Worker	Queen	Drone
Open cell			
Egg	3 days	3 days	3 days
Larva (4 moults)	5 days	5 days	7 days
Sealed cell			
Larva (1 moult)	3 days	2 days	4 days
Pupa (1 moult)	10 days	6 days	10 days
From egg to emergence	21 days	16 days	24 days
Life span after emergence			
Summer bee	6 weeks	3 years	4 months
Winter bee	6 months	3 years	None Present

These times are really important to a beekeeper as they determine how often inspections and other actions need to be taken. Day 8 in the development of the queen bee is of particular importance to beekeepers - see swarming notes.

Eggs

Worker bees build three different sized cells in the wax comb. The queen lays eggs in cells which have been cleaned and prepared by the workers. The size of the cell will determine whether the queen lays fertilised or unfertilised egg. Fertilised eggs to produce females – workers or queens – unfertised eggs to produce male drones.

Comb showing eggs

Larvae

The larvae of the worker and drone honeybees can be seen forming a white, C-shaped and segmented larva at the bottom of the cells. The larvae undergo a series of moults until they are big enough to fill the cell. At this stage they are upright in the cell.

Larvae

Pupae

The fully grown larva forms a cocoon within the cell and the the worker bees will cap the cell with wax. We refer to this as capped brood.

Worker sealed brood

14 Notes for new beekeepers

Drone sealed brood

Food

After the eggs hatch, the larvae of workers and drones are fed initially on royal jelly which rapidly diminishes and is replaced by brood food, and then brood food and honey. Queen larvae are fed only on royal jelly. Royal jelly is produced by workers from special glands in their heads.

Colony Build Up

The colony starts to build up in early spring as the days become warmer and the availability of suitable forage increases. It is quite noticeable that the activities of the queen will be reduced at those times when little or no food is being brought in. Colony number are greatest in July. Swarming usually occurs in May, June or July because the colony is getting very big.

Beehives

There are many different designs of hive. Some people still use straw skeps or clay hives while others use state of the art polystyrene hives. There are advantages and disadvantages with every design. Before you invest in equipment you should really think about what you want to achieve as a beekeeper and what methods suit you best. These notes give you the basic information about hives; you will need to research how the hives are managed and discuss with users the pros and cons before you buy.

The National and WBC Langstroth, Commercial and similar Hives

The National Hive

The National Hive is the most popular hive in the UK. This is an excellent hive for all beekeepers as it is a reasonable size, easy to manage and transport. It also makes life easier for beekeepers to buy colonies on frames or exchange equipment with other beekeepers. Some Beekeepers think the national brood box is too small for a prolific queen. National supers full of honey are easier to handle than supers from other hives.

Most National hives are made from cedar, which do not require any preservatives as cedar has its own preserving oils. This natural oil protects it from the weather and discourages insects. Cedar wood is an ideal timber for hives in the British climate and will last over 15 years naturally so there is no need to paint the hive as this would seal up the grain which will cause mould and condensation problems on the inside. The colony size needs to be carefully monitored during the early spring as a strong colony build up or if the queen has no-where to lay (honey bound) will lead to swarming problems early in the season.

The Deep National Hive
The Deep National Hive is becoming a very popular hive in the UK to allow for prolific queens. It is the same size as the National hive apart from the depth of the brood chamber which allows for deeper frames to be used. The 14"x12" frame greatly increases the total number of cells per frame for the queen to lay in and also for the colony to store greater amounts of pollen and nectar in.

Frames for the deep national hive are called 14" x 12" frames. The frames for supers are the same as in the National hive. This is an excellent hive for all beekeepers as it is a good size, easy to manage and transport.

WBC
Named after the inventor, William Broughton Carr, the WBC is an iconic and highly recognisable beehive design. It is widely used throughout the UK and makes a lovely feature in a garden. It is based on the same principles as the National but with an extra outer wall. This provides the bees with additional insulation and is popular for its looks. The WBC is rarely used commercially because it is costly and inconvenient to use as the outer lifts have to be removed each time for inspection. With a prolific queen who can lay 2000 eggs a day the space available in the brood chamber is considered to be too small.

Langstroth
Named after the Rev. Lorenzo Langstroth based on his observations of the bees and their architecture using the bee-space measurement. Langstroth hives are rectangular and larger than the other designs. It is the most commonly used hive around the world. The large brood box provides lots of space for prolific queens to lay. Commercial beekeepers use both brood box sized and smaller sized boxes to collect honey. If you move to any country outside the UK you'll find these hives. Be aware that Langstroth hives from different countries can be different sizes and are not always interchangeable.

Other Hives popular in the UK

Smith

Popular in Scotland – similar to nationals but the frames have shorter lugs (ends)

Commercial

Commercial hives are exactly the same external dimensions as a National hive, but have a larger brood area due to their internal design.

Dartington, Top Bar and Warre Hives

The Dartington Hive Deep Long Hive

The Dartington Hive is not a common type of hive in the UK as once it is in place it is cumbersome to move with a colony in it. Robin Dartington describes this hive as a break-away from the conventional approach to bee keeping. Focusing instead on understanding the life urges in the colony, centred on the queen, rather than the mechanics of colony behaviour. The management of this type of hive is very similar to a standard hive until the colony is preparing to swarm when the owner just needs to make a few simple adjustments to satisfy the needs of the colony without needing to have on-hand a whole new hive and a complete set of hive equipment ready. This hive uses 14x12 brood frames and standard national super frames.

In recent years the Dartington concept has taken a twist and they are now being aimed more at the urban beekeeper. The Omlet Beehaus (www.omlet.co.uk) is more expensive than a Dartington long hive but I've no doubt that it will last many years longer.

Long Shallow hives have recently become popular. These take normal national brood frames but operate on the same sideways principle as the Dartington.

Top Bar Hives

A top-bar hive is a single-story frameless beehive in which the comb hangs from removable bars. Top-bar hives are usually not portable, and allow for beekeeping methods that interfere very little with the colony. They are popular with some beekeepers who believe it is a more natural form of beekeeping, and with beekeepers with little access to materials and machinery to manufacture complex bee hives that require precise measurements, such as those in developing countries

You won't get a large honey crop but can produce some excellent cut comb.

Warre Hive

A Warre hive has aspects of both a traditional frame hive and a top bar hive but the ultimate goal here is minimal intervention. If your main interest in keeping bees is pollination for your garden then this is an excellent option to look into. It's basically a vertical top bar hive that is meant to mimic the hollow of a tree. It's smaller than a National box and has a box at the top that you fill with wood shavings between two layers of a "quilt" (pieces of cotton fabric) for temperature and moisture control. The entrance is in the front at the bottom but much more narrow than a National so you won't need an entrance reducer. You add additional boxes to the bottom of the hive instead of the top. As the brood area descends you crop honey from the upper boxes.

Disease control is a little more complicated in the Top Bar and Warre type hives than in the National.

The comb honey produced is excellent.

Nucleus Hives

A nucleus hive (often called a 'Nuc' when full of bees) is a small version of the bigger hive. Usually having space for 6 brood frames and often with an integral feeder. These are used as a part of swarm prevention, as a way of producing new colonies or to sell bees.

Extension boxes, queen excluders and super boxes are also available to go on top of nucleus boxes so they can be used to build colonies.

Most beebreeders would now agree that the polystyrene nucleus hive box is the best 'nuc' box you can use. Buy one that goes with your larger hive. The 'poly nuc' is a 'best buy'. I prefer one with a separate top feeder. However wooden boxes are much more robust if you are pushing the box into a thorn hedge when collecting a swarm.

Mating Hives

A mating hive is a very small hive that holds only a few hundred bees and is used to produce newly mated queens from introduced queen cells.

Polystyrene or Wood Hives

Throughout history mankind has changed the design of the containers used to house bees and the materials used to make the 'hive'. We started as 'honey hunters' and have progressed through hollow trees to straw skeps and onto the moveable frame hive. Now we have a variety of 'sustainable beekeeping' hives.

The most significant change over the last few years has been the ready availability of a variety of polystyrene hives – national, langstroth and even top bar hives.

Polystyrene has some advantages over the traditional wooden hives – but they have some disadvantages too. Before you decide which is for you get friendly with local association members and see if they will let you help them look through the different types of hive.

Polystyrene Hives – Advantages

    They are much better insulated
    They are completely waterproof
    Polystyrene does not rot
    They are cheaper than Cedar
    Easy to clean using chemical washes like washing soda solution
    Lighter than wood so not as heavy to lift
    Some are compatible with original wooden hive parts

Polystyrene Hives – disadvantages

                        More easily damaged
                        Need to be weighed or tied down
                        Need to be painted (and repainted at intervals)
                        Parts are less interchangable
                        Need to clean with chemicals instead of heat
                        Bulkier – The outer dimensions can be bulky

If you buy a polystyrene hive don't just buy the cheapest – look at the design. Some encourage the bees to propolise everything so that it is difficult to lift the frames. Some have queen excluders that fit the hive while others do not. The hive should last you 15 years so buy one that is well designed. You can buy feeders designed to work with most polystyrene hives.

The Parts of a Beehive

Most beehives with removable frames have the same basic parts:-

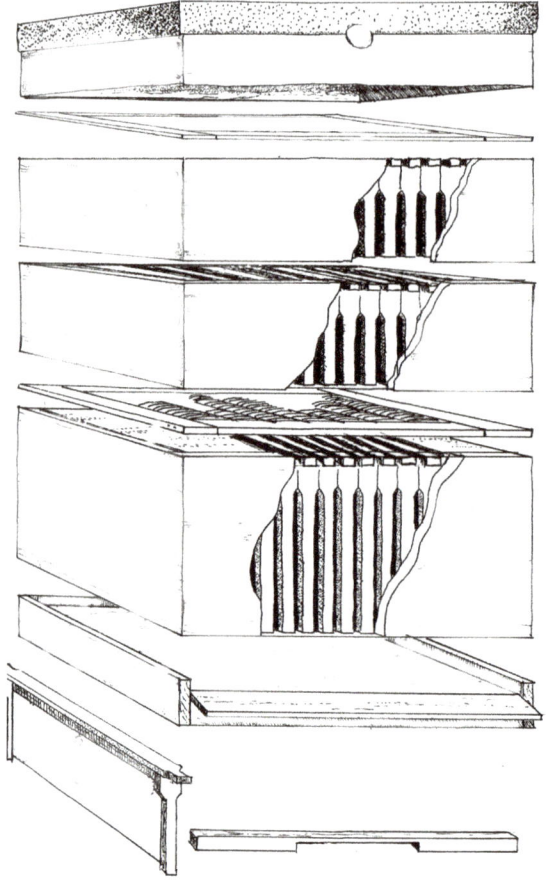

Roof.

Crown board.

Super Box / Honey Box.

Super Box / Honey Box.

Queen excluder.

Brood chamber.

Mesh Floor with varroa insert

A removable entrance block.
Used in wintering to reduced entrance size
Mouse Guard for use in winter

Inside the hive you find 2 types of FRAME:

Brood frames found in the larger BROOD BOX / BROOD CHAMBER

Super frames found in the smaller SUPER or HONEY boxes or chambers
Just to confuse things a super box is sometimes used below the queen
excluder to give extra breeder space - In this case it is called a HALF BROOD BOX.

A frame may contain:
A sheet of Wax Foundation
Drawn Comb - i.e. empty wax comb
Brood Comb - i.e. drawn comb used to lay eggs / larvae / pupae
Honey Comb - i.e. drawn comb used to store honey and/or pollen

Beekeepers will argue forever about which type of hive or frame is best. In an ideal world you should visit an association or individual beekeepers to try out some different types. Which hive you choose will depend on how much you want to spend and why you want to keep bees.

Traditionally floors were made from a solid sheet of wood to help maintain the internal temperatures and keep the frost out. Since the arrival of the varroa mite in this country open wire mesh floors have become the norm to help remove the unwanted mite from the hive. A good size of mesh has holes large enough to allow the varroa to fall through but small enough to keep the hive secure from unwanted pests.

The Entrance Block is fitted to reduce access to the hive during the winter time to help keep the warmth in and unwanted visitors out. During the spring and summer it can be removed when the colony is of a suitable size to defend a larger opening and thus gives the flying bees easier access directly into the hive. The entrance block should be refitted if the hive is being attacked by another colony or if the weather is poor for that time of season.

The Brood Box is the largest chamber of the hive this is where the queen lives and lays eggs. The colony will also store pollen, nectar and honey in this chamber so that the food is within easy reach for feeding the young. The maximum colony size is determined by the size of this chamber which is different depending on the type of hive. During the spring and summer when the colony size has built up beekeepers may split a colony. Some of the frames from the brood chamber which contain eggs, larvae, sealed brood, pollen and honey are moved into another hive nearby. The Queen and some of the frames are left in the original hive. All empty spaces are then filled with new frames with foundation. After a few days the new hive will need to be checked for queens cells - see notes of queen cells. This is a good method to stop the colony from swarming.

Some beekeepers using national hives will combine a brood box and a super box to act as the brood chamber (Brood and a Half) or use two brood boxes to give a bigger volume for brood rearing (Double Brood).

The Queen Excluder Is a thin sheet of either steel or plastic with slots or holes cut in it or set of wires. The best queen excluders are in a frame. The holes are big enough to allow a female worker bees through but too small to allow the slightly larger queen or drones through. Honey super boxes placed above the queen excluder will be filled with honey as the queen is prevented from entering and laying in this area.

The Honey Super is the box of smaller frames in which the worker bees store excess honey. The beekeeper will remove the combs when the honey is capped over and is ready to be extracted. When the weather has been favourable beekeepers will stack 2,3 or even 4 supers on top of the brood box above the queen excluder. The supers are removed at the end of the season to reduce the total space of the hive to just the brood box to help the bees keep warm.

The Crown Board is a sheet of wood/glass/plastic with a hole in the centre and acts as a cover on top of the brood chamber and supers. The board is the internal ceiling of the hive. Most crown boards can also be used as Clearing Boards - they can be fitted with a bee escape(s). A feeder can also be placed over the hole so that bees can be given additional sugar syrup or sugar fondant when the need arises.

Hives should have a metal sheet covered roof, they should be heavy to stop them being blown off in strong winds and help to trap the warmth in the brood box in winter time.

Bee Space

Bees require a certain amount of space in which to move around inside the hive. This is the space between the boxes of frames, between the frames themselves, between the frames and the outside wall of the box, between the top of the frames and the crown board. If the space is not large enough they will fill it with propolis and if it is too large they will fill it with brace comb.

The bee space is 3/8th" or no smaller than 6mm and no larger than 8mm.

As the presence of both propolis and brace comb makes it much more difficult for the beekeeper to carry out regular inspections it is better if you can avoid the problem by using the correct bee space.

The use of well maintained equipment built to the correct dimensions and the use some type of 'spacer' to keep everything in place will help to avoid these problems.

Propolis (bee glue) is made from plant secretions like the sticky stuff on 'sticky buds' and from plant sap. Bees use it to fix things firmly in place. For the beekeeper this makes life difficult and sticky - and bits stop fitting together properly. If you get propolis on anything wash it off with cold water - hot water makes it stickier and spreads it everywhere. It will break off fabric after freezing.

Brace comb is comb built anywhere outside the frames provided. Usually it is built at right angles to the frames and causes them to get stuck together.

Frames, Combs and Foundation

Modern beekeeping is based on the use of removable frames hanging within movable boxes - Brood boxes (where the bees breed their young) or super boxes (where the honey is stored - often called honey boxes).

The national hive, used by most UK beekeepers, can use different combinations of frames

Single Brood Box	11 Deep national brood frames per brood box
	11 Super National honey frames per honey super/box
Brood and a half	11 Deep national brood frames per brood box
	11 Super National frames per brood super
	11 Super National honey frames per honey super/box
Double Brood	11 Deep national brood frames per brood box
	11 Super National honey frames per honey super/box

Some beekeepers use fewer frames (10, 9, 8 or as few as 7) in the honey boxes to increase the yield and to have fewer frames to extract honey from - but you need some experience to do this well.

Frames in both brood boxes and honey supers need to be kept the right distance apart. Most beekeepers use Hoffman self spacing frames in all their hives. This is the simplest option. Spacers keep frames JUST THE RIGHT DISTANCE APART - TWO BEE SPACES APART!

The most common ways of spacing frames the correct distance apart are:-

HoffmanSpacing	'built-in' as a part of the frame (Hoffman frames) DN4/SN4
Castelated spacers	nailed into the brood box or super and the frames fit into them
Metal or plastic 'ends'	fit onto the ends of the frames

Smokers, Tools and Silly Hats

Your local association will advise you on what to buy, when to buy and who to buy from. Gentle bees and a good hive make beekeeping a pleasure. Do not be in a hurry to buy bees. Complete a beginners course to see if beekeeping is really for you - Then spend your money wisely.

Clothing

Well fitting protective clothing and the correct equipment are essential for you to get off to a good start. You will feel safe and secure and so will be more confident.

Hat and Veil - You should always wear a veil to protect your face against bee stings particularly your nose, mouth and eyes.

Bee Suit - Available as jackets or full suits – with a ring-veil or 'astronaut' hood. Get a larger one to cover your clothing and give you room to move. It is worth 'trying before buying' to get one that feels right to you.

Trousers - Wear a pair of over trousers if you have a bee jacket rather than a full suit.

Wellingtons - It is well worth remembering that bees will run upwards and I would recommend that you wear a pair of wellington boots into which you can tuck the bottom of your trousers. It is easier to stop the bees getting into your trousers than to deal with them when they reach the upper levels.

Gloves - Many beekeepers wear a pair of leather gloves and gauntlets which give very good protection but very little 'feeling'. Thick gloves make you clumsy. I recommend long sleeve nitril dispossable gloves – expensive but give good protection and sensitivity. Move over to bare hands when you are more confident.

Tools

Smoker - A smoker is essential to beekeeping. Get a small one for 1 -10 hives. The more expensive the longer it will last. Smoking the bees is an 'art' that requires practise but a 'good smoker' makes beekeeping easier. Use a fuel that burns with a cool smoke, such as dry rotted wood or egg boxes.

Hive Tool - A good hive tool is essential. There are several design variations on the basic chisel and J shaped tools. Which ever you choose the trick is to learn to use the tool well and become a gentle beekeeper.

Soft Brush / Large Feather / Goose Wing

You can remove bees from the top edges of a hive box before replacing a queen excluder/crown board/super by gently encouraging the bees to move away using a soft bristle brush. Personally I never use a brush.

Extractors / Honey Tanks / Ripener /Strainers

Joining a local association will mean that you are able to use association equipment and so avoid buying expensive equipment . You will need to buy lots of jars though - to put all that lovely honey in.

Other Equipment

You can easily spend lots of money on beekeeping equipment. Don't – until you've kept bees long enough to decide what you need.

Smoker

Hive tool held in your hand while you work

Your Apiary

Before you dash out to buy your first colony you need to decide just whereabouts you are going to site your newly acquired bees.

General considerations
1. Is there adequate forage in the surrounding area for the colony to survive?
2. Really importatnt - Can the bees readily obtain water throughout the year?
3. There must be no danger to humans, pets or farm animals. If the bees are sited near animals then the hives must be properly fenced off to prevent the animals from getting too close.
4. Under no circumstances should it be located adjacent to a public footpath even if there is some kind of a barrier e.g. hedge or wall. Bad tempered bees will not be deterred from attacking unsuspecting members of the public even if they have to go over a wall.
5. The site should ensure, as far as it is possible, that the bees will be free from vandalism.
6. The site should be protected from prevailing winds, flooding and not under overhanging trees.
7. The site must be accessible - A full super of honey can weigh as much as 35 lbs!

Detailed considerations
1. The flight path of the bees must avoid footpaths and areas of human or animal activity. You can put them near a high hedge so that they will be forced to fly high but think about what they will be flying into on the other side.
2. There must be plenty of space around the hive(s) for grass cutting, pruning etc. and also for manipulating the colony in comfort.
3. It is advisable to consider future expansion at the outset, allowing space as appropriate.
4. Use a hive stand to lift the hive off the floor and prevent damp and rot occurring and to give yourself a comfortable working height. Don't get 'beekeepers back!'.
5. Hives situated out of sight of neighbours can be advantageous but often making your hobby more obvious will make you a lot of new friends.
6. Swarming can happen even if you are a great beekeeper - Consider their possible landing place when they leave the hive. A nearby bush at a height which is readily accessible for the beekeeper would be useful - but even so, you can't tell the bees where they should go.
7. A nearby shed or other suitable storage place is really useful as you will soon build up lots of equipment that you only use for part of the year.

Three Miles

Use smart phone or computer apps to view the area you live in. Imagine a circle with a 3 mile radius centred on your bee hive. This is the area your bees will find their food in. Look for trees, gardens, parks and other good forage sources. This is also the area your bees could swarm into. Think about schools, playgrounds and similar potential problem areas.

If in doubt about your potential site please do not hesitate to ask an experienced beekeeper to help you look at it.

Selecting an Apiary Site - more Details

It will rarely be possible to find a perfect location for an apiary, but below are some factors to bear in mind when searching for a suitable spot.

Family, neighbours and the public
Unfortunately many people are afraid of bees. While honeybees are usually not aggressive whilst out foraging, sometimes the public confuse wasps with bees and may blame your bees when they get stung. It's good practice to enclose the apiary with a barrier of some sort, such as a hedge or fence to force the bees to fly in above head height. Keeping your hives less visible can also helps reduce the chance of vandalism or theft. Always put up a sign near your hives warning that honeybees can be dangerous.

Bees and horses do not mix well. Keep your hives away from horse stables or bridleways. Bees need to collect water in the summer and sometimes they like collecting the sweat from the back of horses. This can lead to the horse and rider getting irritated.

Beekeepers can reduce any potential problems with the public by keeping good tempered bees and replacing the queen in any bad tempered colonies. This will also make inspecting more pleasurable for the beekeeper.

Forage
Find out the amount and type of food sources available around your potential site, by taking a walk, by asking local beekeepers and by looking on GoggleEarth. . The number of hives that an area can support is related to the amount of available forage. Often ten colonies, with all the work entailed, may well only bring in the same amount of honey as five colonies if the forage is limited.

Remember that you are concerned with forage from April through to the end of September. Think about sources of early and late supplies of pollen as well as the main nectar plants. The bees need pollen continuously whilst rearing brood, which in the UK is usually from February through to late September or even October/November if we have a warm autumn. Colonies only store about 1kg of pollen (about one week's worth of their requirements), during summer – so a pollen shortage will quickly impact on the health and development of brood emerging 2-3 weeks after the shortage.

Bees usually forage within a 2-3 mile radius of their hives. It takes 8 Kg of nectar evaporated down to produce 1 Kg of honey; it takes about a dozen bees to gather enough nectar to make just one teaspoon of honey, and each of those dozen bees needs to visit more than 2,600 flowers. So there will be a limit on how much forage is out there – no location supports an infinite amount of colonies. The density of bee forage in most areas will not support more than ten to fifteen colonies in one place.

Seasonal Out Apiaries
In an urban area gardens and parks usually provide a variety of forage throughout the year. The bees can be taken to a temporary apiary site in the countryside to exploit a seasonal crop such as heather or oil-seed rape. Bees are usually moved to these out-apiaries for a few weeks. Prior permission needs to be obtained from the landowner. The traditional rent for an apiary is 2 jars of honey per hive annually.

Seasonal beekeepers should take into account walkers, pony trekkers, shooting parties etc and place the hives away from footpaths. Avoid having hives near main roads too, as bees can be hit by passing cars as they attempt to fly over the road.

The Apiary Environment

- A flat site is easier to place hives on!
- South facing is warmest.
- The site should be sheltered from wind, so that foragers don't struggle to land at the hive entrance and the roof stays on. A hedge provides good cover, as the small amount of wind coming through prevents areas of turbulence which occur behind a more solid surface such as a wall.
- It should be a site which does not flood. Hives on moorland have been partly submerged and even washed away after torrential downpours. Generally avoid muddy sites or low-lying areas near rivers.
- Keep hives away from the bottom of dips in the land as these are likely to be frost pockets and therefore a few degrees lower in temperature. Bees won't start foraging until the temperature immediately outside the hive warms up enough (12-14°C for nectar foraging).
- Most books advise that sites under trees are unsuitable because they are usually damp. Bees naturally produce water vapour as part of their metabolic processes. Excess moisture is usually removed by bees standing at the entrance and fanning, but if the location is too damp, they may not be able to sufficiently reduce the humidity and mould may start to grow on the woodwork and pollen stores. However I keep a lot of colonies under light tree cover with no problems.
- Dense foliage cover can make hives too wet and cold; however some shade in the afternoon helps prevent bees having to work hard to cool the hive or even dying from heat exhaustion or collapsing honey combs. Position your hives where they will be woken up by morning sunshine but shaded during the hottest part of the day.
- If your hives are in a rural location, fence them off from livestock like cows or horses which might like to use them as a scratching post and knock them over.

- The bees will need a water source to produce brood food, dilute honey stores and cool the hive in hot weather. If a suitable pond or stream is not available consider providing a shallow water source in a sunny position with stones bees can rest on to avoid drowning. Place this away from their main flight paths to avoid fouling. Adding a distinctive smell, such as peppermint essence, will help the bees find the water.

Access

Easy access to a site throughout the year, with a hard path down to the apiary, is important. Honey supers are heavy, so if you are using an out apiary it helps if you can park your car nearby.

Space

You need room to stand while inspecting and somewhere to put the roof and supers down. Generally you should allow nine times the hive footprint area per hive, though this is not a strict rule as two hives placed side by side reduce the overall space needed.

Make sure you have enough room to add a hive or two. Swarming means two colonies can quickly become four during a single season. Even if you plan to recombine hives following an artificial swarm you will need extra space temporarily

A small community apiary

BEES AND NEIGHBOURS

Generally you will be able to get on well with your neighbours if you talk to them about your beekeeping and give them some honey.

There is no doubt that many people are genuinely afraid of bees. This is not necessarily because of the possibility of stings but is a real fear of what they regard as "creepy crawlies'.
A massive swarm in flight can induce panic. Fear nearly always gives rise to feelings of anger and aggression. Beekeepers need to be sympathetic towards these feelings and understand that bees can be regarded with dread by a some people.

It is important that we keep bees in such a manner as to cause the least possible annoyance and distress to our neighbours. Failing to do so brings the craft into disrepute and negates your insurance.

Our Legal Responsibility
If bees are kept in large numbers, or are sited inconveniently to neighbours, or are bad tempered, or are poorly managed so that a neighbour is prevented from leading an ordinary life in his/her house or garden then a state of nuisance can arise. Legal resort to the Courts may ensue resulting in an injunction to prevent keeping bees on that particular site and an award of damages.

The nuisance may not be merely stinging. Faeces spotting washing, danger to children and pets, the frequent occurrence of swarms are all factors. The bees do not have to be a public nuisance for a case to arise.

What to do - and not to do

A badly managed colony of bees is potentially dangerous and it is clearly antisocial to keep bees in such a way that neighbours are made uncomfortable or are intimidated. It is therefore a first priority for new beekeepers to learn to handle bees efficiently - if possible before he/she acquires bees of their own. Local Associations run training courses so that you can learn good practice before you get your own bees.

- Do not crowd a large number of hives into a small suburban garden.
- Stacks of empty equipment can create a negative impression.
- Pay careful attention to the siting of hives particularly proximity to boundary fences.
- Watch the flight path. Erect some sort of screen to get the bees up to cruising height quickly.
- Do not tolerate bad tempered bees - requeen the colony.
- Feed late in the evening and feed all colonies at the same time. Avoid spillage.
- No swarm control system is 100% effective but do use a method to reduce the incidence of swarms to an absolute minimum.
- If possible inspect your bees between 11 am and 4 pm and in good weather conditions. This is when bees are busiest and least aggressive. If you inspect in the evening or at the weekend try to avoid times when neighbours are enjoying their gardens.
- Provide a source of clean, slightly salted water early in the season so that the bees become accustomed to using it. Keep it constantly topped up.
- Try to involve your neighbour's interest in your beekeeping. It is sometimes possible to do this through children, most of whom will be interested in a peep into a hive. Keep a spare veil and perhaps an old pair of gloves.

If Things Go Wrong

It is always best to take positive action to prevent trouble before it happens and a gift of honey occasionally will help to keep relationships positive.

Avoid aggressive encounters. Think carefully. Is it really your fault? If it is then the only course open is to remedy matters. Perhaps offering to reduce colony numbers or planting /constructing flight barriers to keep the bees above head height.

If you think the complainant is unreasonable you should still try to smooth things over. Your Association Secretary or Chairman can confirm that you are keeping your bees in a proper manner.

If you cannot satisfy your neighbour and they are resolved to pursue their demands through a solicitor or in the Courts, two options are open to you. You can reduce the number of colonies you keep (or move them to an out apiary) or you decide to defend the action. Consult your association for support.

Starting Beekeeping

Every beekeeper will give you slightly different advice about 'starting up'. The notes below with give you some indication of what you will need to 'become a beekeeper' over a two year start-up period.
You can cost this out easily enough visiting any of the many websites selling beekeeping equipment.

The Best Time to get Bees

As a new beekeeper you should aim to get your first colony of bees in the period late May to early July, either as a swarm, a nucleus or a small colony. You would treat each of these slightly differently but this would give you enough time to build the colony up for winter. If the weather is good you might well go on to produce a surplus of honey in the first summer.

The beginner is best advised to purchase a strong 5/6 frame nucleus containing a ready mated queen who has already started laying. She should be marked.

Minimum Start-up Equipment

In the first summer you would need to have the following (of which ever style you choose)

Somewhere to keep two hives within a metre of each other providing a clear flight path away from neighbours.

For the Bees
1 double or 2 single hive stand(s)
A mesh floor with varroa insert and removable entrance block and mouse guard
1 brood box with enough frames to fill the box when the bees are installed and enough wired foundation for these frames
1 Queen excluder - wire or slotted but preferably on a 'frame'
2 'super' honey boxes with a full complement of frames and foundation
1 Crown/Clearing Board
Bee escapes to fit your clearing board - usually 'porter escapes' at first
1 rapid feeder - the larger the better
1 Record Card
1 Deep Roof
Varroa Treatment of your choice

For Yourself
1 pair of wellingtons or equivalent
Overall and veil or all in one beesuit or bee jacket & trousers
Gloves to suit yourself
1 smoker
1 small garden hand sprayer of sugar water
1 hive tool

For Honey Processing   Borrow everything you might need from the association in the first years

You can spend from a hundred pounds to thousands of pounds on equipment. Don't buy it until you're sure you need it.

When you get your bees build them up in the brood box and add supers as the colony develops during the first year. You'll need lots of help and advice this year so make sure you ask. Exactly how your colony will develop depends on what you started with, the weather and what plants grow near you. You can only take one step at a time.

If you started with a swarm on foundation put a feeder on 24 hours after hiving it and keep feeding until they have drawn most of the foundation. You can continue feeding until a super of foundation has been drawn out. Reduce the entrance while feeding. In a prime swarm with a mated queen laying will commence within a few days. A virgin queen may not start laying for four weeks.

If you started with a nucleus then transfer it to a brood box and encourage them to draw out the extra combs. A strong nucleus with a laying queen may need supers added within a couple of weeks.

If you start with a small colony build it up with regular small amounts of feed but keep the entrance small. Tender nursing this year will build a good colony for next year

Purchasing Bees

Bees should only be purchased during the flying season, preferably in Spring. Movement during the winter could cause the clustered colony to die out. In the summer you can inspect the bees you buy.

The best course of action is to purchase a NUCLEUS (Nuc) from a recognised supplier. This is a 'mini' colony which you can build up as your first year of beekeeping progresses. You will need to purchase a complete hive to put your bees in. Get advice from an association member.

There are leaflets, available from the National Bee Unit BEEBASE (NBU) and BBKA websites, that gives a detailed description of what constitutes a good quality nucleus of bees. Read these and check that what you buy fulfills all the criteria listed.

Get a colony from somebody local who you know and trust.

Honeybees will forage up to 3 miles from their hive and this must be borne in mind when purchasing bees. If they are used to foraging in your area there is a good chance that when you move them into your garden they will return to their original site rather than into the hive in your garden. So bees must be purchased from someone who is at least three miles away from you (as the bee flies) or who has the facility to move them away from his apiary and move them back to your garden at a later date.

Bees are best moved at night when they are less prone to fly and when it is comparatively cool outside. The entrance should be fully closed with a foam strip and the roof removed and replaced with a travelling screen. The whole hive should then be securely strapped together to prevent movement in transit.

Having safely transported the bees to their new home put the hive down on its new site. Remove the strapping and leave the colony alone for half an hour to settle down. Then gently remove the foam strip in the entrance and, provided the bees are not too agitated, the travelling screen. Replace the crown board and roof and leave them alone.

Do not be tempted to inspect them too soon – They are best left alone for a few days to help them to adjust to their new surroundings.

Beekeeper

How to transfer your bees from a 6 - frame nucleus hive into a full hive

Here is what you will need to transfer the bees into a beehive.
Hive floor and entrance block.
Brood box
Five deep frames and foundation, or four if using a WBC hive.
Frames should be DN1 with narrow spacers/castellated spacers or DN4 Hoffman self spacers
Crown board.
Empty super to hold feeder
Feeder
Roof

A Note about frame spacing

The nuc frames may not be Hoffman, nor will they always have spacers already on the frames – so you may need to have some spacers to put onto the supplied brood frames from the nuc colony. Ask in advance.

Collecting the nuc to take home

1. Collect your nucleus hive of bees in the evening, when the bees have stopped flying. The supplier will usually seal the nuc box the night before you collect the bees.
2. Place nuc hive in position on the hive stand.
3. Open the entrance.
4. Leave bees to fly and to settle down for a day or two.

Decision: Do you want to run your hive the warm way or the cold way?

Warm way, with the frames parallel to the entrance allows you to work the hive easily from behind. The cold way, with frames at right angles to the entrance, requires you to twist if working from behind but not if you work at the side of the hive.

Putting the nuc frames into a hive

1. Choose a warm sunny day when the bees are flying well.
2. Move nuc hive to one side and place your new hive on the stand with the entrance block in place to reduce the entrance.
3. Place three frames of foundation, with spacers if needed, in one side of the brood box.
4. Open nuc hive and gently smoke your bees.
5. Lift out first frame of bees, add spacers if needed, and place in hive.
6. Continue to transfer frames of bees making sure that they are kept in the same order and orientation as they were in the nuc box.
7. Add the last two frames of foundation with spacers to the brood box.
8. Shake any bees still in the nuc hive into the new hive.
9. Fit crown board.
10. Fill in a new hive record card.
11. Fit roof.
12. The same evening, give the bees a feeder of sugar syrup. Buy this ready made or make up your own - 1Kg sugar dissolved in 500ml water.
13. Continue feeding until the bees have drawn out all the frames of foundation. If possible feed a little each day rather than give them large quantities in one go - the ready availability of food might encourage swarming.

Examining Your Bees – Inspecting the Colony

APRIL is the start of regular beekeeping activity when honeybee colonies change from simply surviving the winter months to developing into active mature colonies. They go on to reproduce by swarming, replacing an ageing queen and gather enough food reserves to survive through the next winter. In the process, with our help, they may produce surplus honey for us to gather. As always, the weather affects how quickly colonies develop and therefore how and when we act to help them.

The queen's egg laying accelerates as the weather and available forage improves. If there is an early warm spring you may see eggs laid in drone cells and the first swarms may appear before the end of April. All the incoming nectar and pollen is used for colony maintenance. So from late April you need to check the colony to make sure that all is well and to look out for signs of swarming.

The brood population may be increasing faster than the adult population so if there is a sudden cold spell, there's a risk of brood starving and chilling as the population of adult bees may be too small to feed and cover the brood. You can help by feeding the bees during the poor weather.

April to June is probably the busiest time helping your bees to develop into productive colonies. This entails opening the hive to examine and assess the condition of the colony regularly. You need to check that it has adequate reserves of food, that the bees have enough space, they are healthy and whether they are preparing to swarm.

How often we carry out the inspection is determined by the timetable for swarming. A swarm usually leaves the hive after the first queen cell is sealed which is eight days after the egg was laid. So if the egg was laid in the queen cell as soon as you closed up the hive after one of your regular inspections the swarm will leave in another eight or nine days. Sometimes we may have to miss a day because of the weather or work and most of us live to a seven-day cycle, so the best plan is carry out a regular inspection every seven days. This will give a day's leeway for unforeseen circumstances or bad weather.

There's quite a lot to find out in the short time the hive is open during an inspection. Use these questions as a guide to what you should be looking for.

1) Does the Queen have enough room?
The queen needs plenty of empty cells where the she can lay eggs. Sometimes you find that the brood combs are full of food. You should replace some of these with drawn comb or foundation. The combs you have removed should be stored in a bee-proof box or spare brood chamber. They will be useful for feeding a colony or nucleus hive later on.

2) Do the workers have enough room?
The workers need enough space to store nectar. The queen excluder and first super should be put on when seven or eight brood frames contain brood. More supers may be added as the bees cover 75% of the frames in the super. In good weather more than one super can be added at the same time.

3) Is the queen present and laying?
There is no need to try to find the queen. If you can see eggs then that is evidence that the queen was in the hive up to three days before. You can expect the rate of egg-laying to increase week by week from February to July.

4) Is the colony steadily increasing in size?
You can measure colony growth by counting the number of frames containing brood. Do this every week and record the results. If you have more than one colony you can compare their progress. If you have only one colony find out how fellow beekeepers' colonies compare.

In late April some colonies will have seven or more frames with brood, most will have five or six and a few will be weaker. If the colony is not developing steadily from April into May then you may need to put a new queen into the colony.

5) Are there any queen cells?

Most healthy colonies build a number of potential queen cells called 'play cups' or 'queen cell cups'. They are not significant while they are empty. There is no point in removing them as the bees will simply build more but it is important to check each one carefully.

If you find a queen cell containing a larva in a pool of milky brood food then this means that a new queen is being raised and swarm control action is needed. The cell will be sealed eight days after the egg was laid or five days after the egg hatched. If you find a larva and food in an open queen cell you need to take swarm control action straight away.

Finding a sealed queen cell means that the swarm and the queen have probably already flown.

6) Is there enough food for the colony ?

You should always make sure at every inspection that the bees have a reserve of food of about 4 kg (or the equivalent of two full National brood frames) in case the weather turns cold or wet and the bees are unable to forage. The reserve will cover their needs until the next inspection. If they need food you can feed sugar syrup in a feeder.

7) Are there any signs of disease?

Learn to recognise healthy brood and look out for and investigate anything abnormal. Learn to recognise the signs of varroa infestation, EFB and AFB as these are notifiable diseases. If in doubt ask an experienced beekeeper. The National Bee Unit at APHA have excellent publications that are available to download free at http://beebase.csl.gov.uk/

It is difficult to remember afterwards what you found out during your hive inspection so keep a record of the state of the colony after each inspection. Record keeping can be a simple note in an exercise book, filling in a record card, notes kept on a computer, using the BEEBASE record system or you can use you smart phone. There are a host of apps available.

It is useful to record the following:

Date

The number of frames in the brood box occupied by bees

Number of frames with brood

| Brood stage | Brood area | Brood pattern | Stores | Gentleness |
| Slow moving | Queen seen | Queen cells | Weather | Disease |

You can add brief pros comments as needed.

An example of a Hive Record Card useful for beginners is included at the end of these notes. See record card example on page 112.

Supering (Adding Supers)

It is a good plan to add more supers before the bees need more space. I suggest that when the bees are occupying 75% of the frames in the first super you put on in April, it is time to add another super. It does not matter if the super is not yet being used to store nectar. The increasing population of bees needs space to park themselves as well as to store nectar.

Nectar takes up three times more space to store than the honey to which it will be converted. So this month be generous in adding supers well ahead of the bees' needs or else congestion in the brood chamber may trigger preparations for swarming.

When you add the second and subsequent supers with foundation it is a good idea to add a couple of frames with drawn comb (from the first super) in the middle of the box to encourage the bees to go up into the super. Place it on the queen excluder underneath the first super you added. This is known as 'bottom' supering. It puts the foundation near to the brood chamber where the warmth helps wax production. Do not risk contaminating the honey with sugar syrup by feeding when nectar is being stored. If the additional super is full of drawn comb it does not seem to make any difference if it is added underneath or above (top supering) the existing supers. If you have some drawn comb and foundation you can alternate the frames in the super - often referred to as checker-boarding.

Now it's up to the bees and the weather to make the most of the foraging opportunity in your area to fill the supers with honey.

The 2 Hive System

In spring your first colony will build up as the weather improves. Unless the weather is very good don't open the colony until early April. From the first inspection you should aim to check your bees regularly as the weather permits at first but from May onwards every 7 days.

As the colony builds and the weather improves add a super or two. Drawn comb is best but add foundation if this is all you have. When the bees have filled 7 frames with nectar add another super. Stop adding supers in late July. If you live in a oilseed rape area you will need to extract honey as it is prepared for capping by the bees otherwise you will have to use unwired foundation and melt the honey out at the end of the season.

If all goes well you could end up with 3 or 4 supers full of honey and no signs of swarming… and no use for all the extra equipment.

If the bees start to make queen cells and show signs of preparing to swarm then you can artificially swarm the colony by any method you fancy or you can do a simple split making sure that half the colony (5 brood frames) with the queen is placed on the original hive site and half the colony (6 brood frames) with the queen cells (reduced to two good cells) is within 1 metre of the other hive. Give this second colony any combs of stores.

Fill the remaining space in the brood boxes with frames of foundation. Divide and honey supers between the colonies but give the colony with the queen a super of foundation as well.

Continue to inspect the colony with the queen at regular intervals.

Leave the colony with the queen cells alone for 3 weeks before inspecting. Check for eggs. Build this colony up as last year. Mark the queen using a crown of thorns cage and marker pen. Remember the colour code:-

Will You Raise Good Bees

Year with a 1 or a 6	Will	=	WHITE
Year with a 2 or a 7	You	=	Yellow
Year with a 3 or a 8	Raise	=	Red
Year with a 4 or 9	Good	=	Green
Year with a 5 or 0	Bees	=	Blue

In Autumn

After you have harvested any honey the two colonies need to be joined together.

Go through the original colony. Kill the queen. If you can't find her to do this then don't worry - just carry on as below. When the bees have stopped flying seal the original colony with foam. Move to one side. Move the colony with the young queen to the original site. Take off the crown board and cover the top of the hive with two sheets of newspaper. Make some small slits in this.
Lift the original colony off its floor and place the brood box on top of the paper. Add roof.

The bees in the top box will eat through the paper and mix with the other bees. Feed the colony well. Over winter as a double brood chamber. In the following year artificially swarm the bees by separating the two brood chambers in a similar manner to the system described above.

Please note that bees do not always do as instructed so we cannot guarantee that any system will work but in theory this should. Good luck. And remember, if in doubt, ask .

Queen cells and Queen Cups

When inspecting your bees you might see:-

Queen Cups / Play Cups
These are the start of queen cells that will probably never be completed. If they are empty then you can ignore them or remove them.

Open Queen cells
An elongated cell hanging down between the combs – open at the end so you can look into it and see the larva.
The larva is being feed royal jelly. The old queen will still be present unless this is a supersedure cell.
There should still be eggs and young larvae present in the combs.

A Sealed Queen cells.
An elongated cell hanging down between the combs – closed at the end so you can't look into it.
The old queen leaves with the swarm once the cell is sealed.
If the queen cell is more than 11 days old there will not be any eggs in the comb.

An empty Queen Cell
A dark coloured elongated cell hanging down between the combs with an open end.
The cell will be empty.
A virgin queen has emerged.

Queen cells and Queen Cups

The colony builds queen cells for three reasons:-
Queen cells are produced by the colony when the pheromones produced by the queen are at low levels. Usually because either the old queen is failing (supersedure) or the colony has become very populous (swarming).

Replacement
The old queen is dead or injured.

Supersedure is the changing of the queen in a colony without swarming taking place. In supersedure there is usually only one queen cell produced usually placed in the centre of a frame.

Swarming
When the bees intend to swarm there will be several queen cells produced, usually on the top, sides or bottom of the frame. The colony reproduces by dividing to form original colony and a swarm(s). When the worker bees decide the colony is large enough and external conditions are favourable they will initiate swarming The workers stop feeding the mature queen so that she stops egg laying and contracts her abdomen so that she can fly. The first prime swarm that leaves the hive will consist of around 40% of the bees in the hive. Young bees with the ability to build comb are an essential part of the swarm. All of the bees gorge themselves on honey in order to have the food stores needed to build a new home. The experienced flyers act as scouts. Swarming bees are unlikely to sting.

Do Not Destroy Queen Cells without thinking first
Managing swarming is important as it usually happens at a time when there is a strong honey flow, and if successfully controlled it will have the benefit of considerably increasing honey production. If you do find a queen cell don't destroy it until you've inspected the whole colony. When you've decided whether supersedure or swarming is occurring then you can take action.

An essential factor in controlling the swarming colony is to carry out examinations of the colony every seven days. You can try to prevent swarming or you can try to control swarming

Swarm Prevention

Swarming is a natural process that enables the bee colony to reproduce. A healthy thriving colony may swarm in the queen's first year.. Most colonies will swarm in their second or third year. We get notice that a colony is likely to swarm when we see eggs or larvae with food in developing queen cells.

Queen cell production is mainly triggered by the diminishing queen substance reaching each worker. The queen's genes and age play a part. Some bees are more inclined to swarm than others and the quantity of queen substance a queen produces declines after she reaches 18 months of age.

Other factors include the availability of early crops such as oilseed rape and good weather, which lead to population growth. This means colonies can become congested which interferes with the distribution of queen substance.

We cannot do anything about the weather but beekeepers can prevent congestion in the colony by regularly checking if the colony has room to expand and by providing extra space for the colony.

Give the queen more comb to lay eggs in by:-
Increasing the size of the brood chamber by adding a brood super or second brood box.
Replacing brood combs filled with honey with fresh comb.
Changing old dark comb with fresh comb

Give the colony more room to store honey by:-
Adding more supers (honey boxes)
Removing some frames filled with sealed honey stores
Extracting the honey from full supers and placing them back on the hive

If you find that you have a 'swarmy' queen you can plan to replace their queen with one that has been raised from a 'non-swarmy' stock.

You can reduce the occurrence of swarming by replacing any two-year-old queens with young ones.
You can buy queens but the best practice is to learn to raise your own.

What to do if you find a queen cell

Examine the Colony	You find unsealed and sealed Queen cells
Ask	Are there eggs and very young larvae present?

Yes	Your queen is probably still in the colony
	You need to carry swarm control or make an artificial swarm:-
	    Colony Split
	    Dameree Method
	    Pagden Method
	    Snelgrove Board Method
	    Horsley Board Method
	    And variations on a theme.

No	Select the frame with the nicest looking queen cell – mark this frame Destroy all other queen cells Close the hive 3 days later go through the hive again – remove any new queen cells Leave the hive for 2 or 3 weeks before inspecting again The virgin should now be mated and laying

Swarm Control

As soon as you find occupied queen cells you should adopt your chosen method to manage the swarming procedure so that you don't lose half of your bees over the horizon and probably most of your honey crop for the year.

There are many different procedures you can use. If you have started with a particular method learned from an experienced beekeeper~ persevere with it so that you become confident with one method before you try another. Don't try to follow bits of different methods.

All methods of swarm control are based on the idea that a colony is made up of 5 parts:
The queen
The queen cells
The flying bees
The brood
The young bees

By separating some of these components we can mimic what the bees do when they swarm.

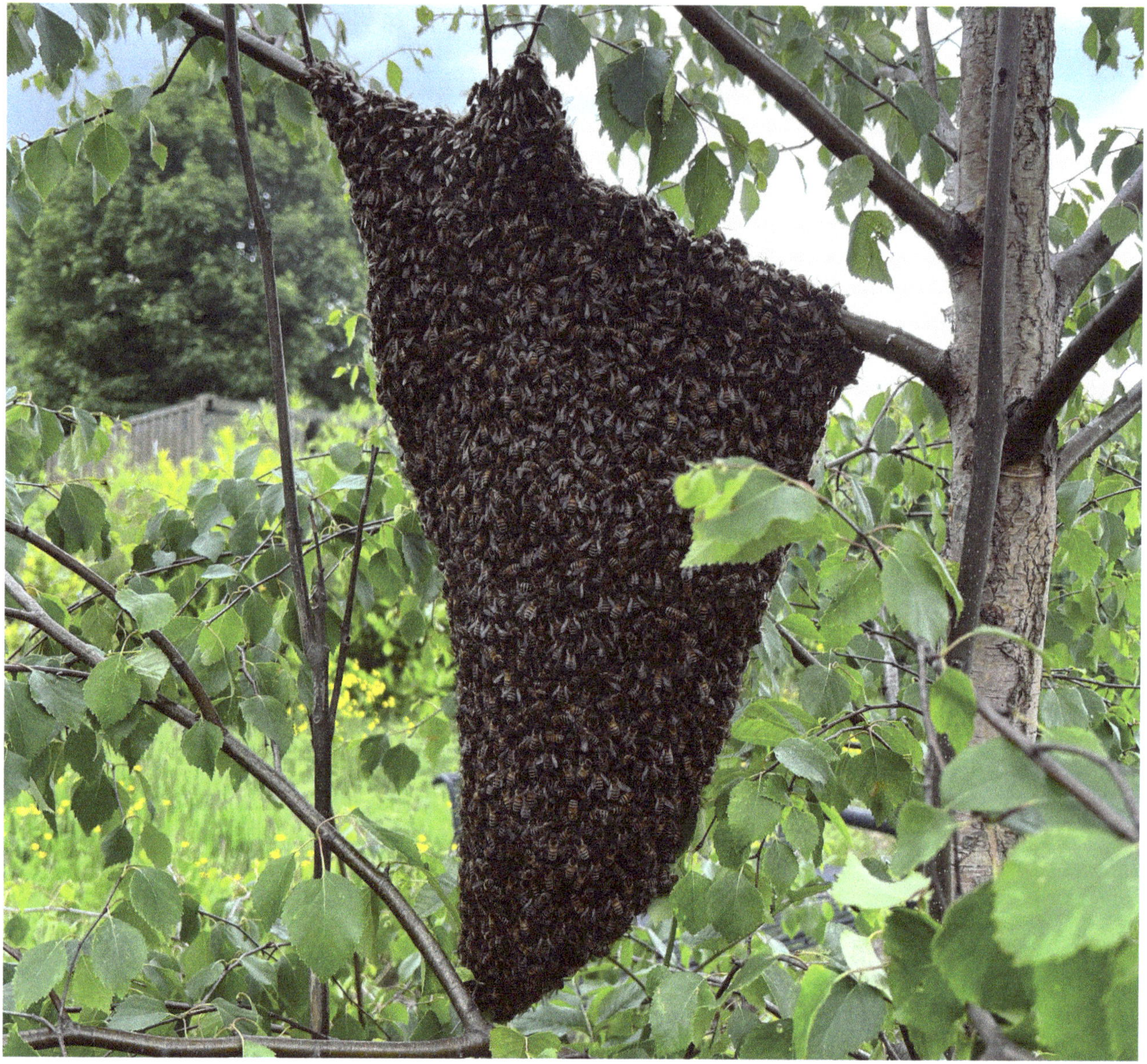

A Nucleus Colony

The term 'nuc' is used to refer to either a small brood box with 3, 5 or 6 frames and a small entrance or the colony of bees kept in the small box. A good nuc box has an integrated feeder or you can use a frame feeder.

Nucs are usually made up from strong colonies. The best nuc will consist of 3 frames of mostly sealed brood and two frames with pollen and honey plus an additional frame. All the frames should be covered with bees. Additional young bees can be added to the nuc from 2 more frames. Pick up the frame and shake it once over the old colony to get the flying off, then shake over the nuc box to remove the young bees. The nuc must also have a queen or queen cell so it can grow and develop into a full colony.

The NUC method of Swarm Control

If you find unsealed queen cells.
1. Find the queen. Put the frame with the queen on into the nuc box. Make up the nuc to the full strength as described above. Remove any queen cells from all of these frames. Remove the nuc to a new site or stuff the entrance with green grass and place in the apiary – near the original hive if you want to unite the colonies at a later date.

   This colony can quickly be built up into a full colony as the old queen will return to lay very quickly. The many young bees will draw out comb. Move them to a full brood box with drawn comb or foundation and feed them to build the colony up quickly. Use a small entrance and dummy boards which you can move apart as the colony grows.

2. Select 2 or 3 unsealed queen cells which look as though they will produce nice big viable queens. Mark the frames with these on with a drawing pin. Brush or gently shake off bees to check for queen cells. Remove all other queen cells.

3. Close up the original hive and leave for 7/8 days.

4. Reduce the number of queen cells to 1. You could make up more nucs with the other cells if the colony is strong enough. Brush or gently shake off bees to check for queen cells.

5. The original, still strong, colony with the emerging virgin can be left for at least 3 weeks without an inspection. By this time she should be staring to lay. Super as necessary – usually as soon as she starts laying.

The Pagden Method of Artificial Swarming

This method will maintain the maximum foraging force with the queen who resumes egg laying without interruption which therefore minimises the loss of honey production. If you follow the procedure carefully and do not miss seeing any queen cells it (nearly) always works.

You will need a spare hive comprising brood box, floor, crown board, roof and a full complement of frames and drawn combs, if you have them, or foundation.

As soon as you see open queen cells containing larvae you should create an artificial swarm. Do not shake any bees off the combs at this stage because this is likely to injure the developing queen. Instead, gently push the bees aside with your fingers so you can see the whole of the comb and especially around the edges to find queen cells.
You may gently brush the bees off the comb if you want to remove all the bees, again taking care of the queen cells. Complete your initial examination of each comb and note which frames carry queen cells.

You will have already taken off the roof, supers and queen excluder. Now proceed as follows:

1. Move the original floor and brood box (with brood, bees and queen) to hive stand about 1 m to one side of the original hive stand, with the entrance facing in the same direction.

2. Place the new brood box with frames of drawn comb or foundation where the original hive was placed. Remove a frame of foundation from the centre of of this box and keep it for later.

3. Examine the original colony.
   Find and remove the frame with the queen and a small area of brood in all stages, but no queen cells. and place it in the middle of the new brood box. Place any available drawn comb on both sides of the introduced frame so that there is room for the queen to resume laying straightaway.

4. Check for food reserves on combs in the original colony and add the spare frame of foundation to this box. Add the crown board and roof. Reduce the number of open queen cells to one or two. If necessary, feed sugar syrup two days later.

   All the flying bees will return to the original site and enter the new brood box.

5. Put the queen excluder and any supers onto the new brood box. You have now created an artificial swarm comprising the old queen, the bees in the honey supers and the flying bees.

After seven days (that is at least one day before the first virgin queen is due to emerge):

6. Move the original brood box to the opposite side of the original site. The bees that have become foragers during the week will now return and, finding there is no hive there, will return to the nearest hive which is in the new brood box on the original site.
    This reinforces the foraging force in the original colony and will also eliminate the risk of a cast swarms leaving with a new queen because there will now be fewer bees in the original colony.

7. Examine the new colony to check that the old queen has resumed laying and that there are no new queen cells.

    After another 14 to 21 days:

8. Check that the new queen in the original colony has mated and has started to lay.

In autumn you can unite the two colonies, first of all removing the old queen.
Alternatively, if you wish to increase the number of hives you own, you can prepare both colonies for the winter.

The Pagden Method

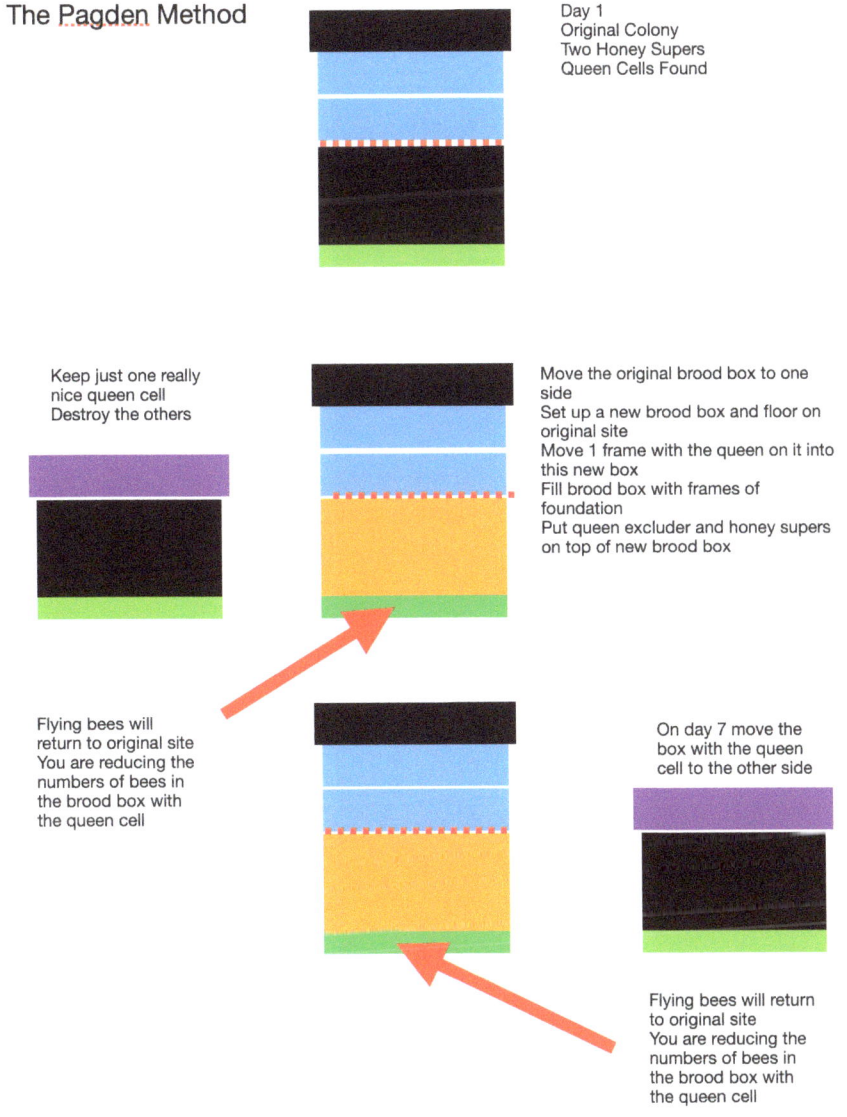

Day 1
Original Colony
Two Honey Supers
Queen Cells Found

Keep just one really nice queen cell
Destroy the others

Move the original brood box to one side
Set up a new brood box and floor on original site
Move 1 frame with the queen on it into this new box
Fill brood box with frames of foundation
Put queen excluder and honey supers on top of new brood box

Flying bees will return to original site
You are reducing the numbers of bees in the brood box with the queen cell

On day 7 move the box with the queen cell to the other side

Flying bees will return to original site
You are reducing the numbers of bees in the brood box with the queen cell

Pagden method

If you can't find the Queen

It can sometimes be difficult to find a queen in a large colony, especially if she is unmarked, because there are so many bees in the hive. If you find queen cells but can't find the queen you can still control swarming by creating an artificial swarm.

1. As you examine the combs as in stage 3 above and select a comb with some brood in all stages but no queen cells and place it in the spare box on the original site.

2. Now carefully brush all the bees off all the combs into the new box on the original site, replacing all the combs, some with brood but no bees, into the original brood box.

3. When you are sure that all the bees, including the queen, are in the new box on the original site, replace its queen excluder and supers. Place the brood box with combs, open queen cells and no bees, on top of the supers. Replace the crown board and roof and leave it overnight. This will allow time for the nurse bees to move up to cover the brood and keep it warm.

4. Next day, lift the brood box containing combs of brood and nurse bees on to a floor about 1 m to one side of the original site. You can now replace the crown boards and roofs on both hives and proceed from Item 6 above.

Finding a Sealed Queen Cell - Your Bees have swarmed

You need to look carefully for occupied queen cells in the brood box during your regular seven-day examinations so that you do not lose half of your bees in a swarm. Even experienced beekeepers can miss seeing a queen cell during a hive inspection.

If a sealed queen cell is found it means that the first prime swarm probably left the hive on the day or day after the first queen cell was sealed. The bees sometimes delay their departure because of bad weather.

When inspecting the hive you will need to ask: Has a swarm already left? Has a virgin queen emerged? You need to know the answers before you decide how to prevent the loss of more bees in a secondary swarm or cast.

First look for eggs or young larvae at the bottom of the brood cells. Larvae that make a well-formed 'c' are more than two days old. If there are no eggs or young larvae it's likely that the swarm has already flown off with the old queen.

Mark any frame with a queen cell on with a drawing pin or in some other way. The best queen cells are surrounded by worker brood and contain a well developed pearly white larva in an ample bed of royal jelly.

Next look carefully at sealed queen cells to find out if a virgin queen has emerged. You can check sealed queen cells by gently touching the tip with your hive tool to see if the hinged cap has been closed after the queen emerged. The workers do this quite often, sometimes enclosing a worker bee scavenging the remains of the royal jelly inside.

If you find that a virgin queen has emerged, she may be in the hive or she could have left with a secondary cast swarm. In this case it is likely that there are other sealed cells with virgins ready to emerge. You can open all the sealed queen cells with the tip of your hive tool, releasing one or more virgin queens into the colony.

Having released a virgin queen you should destroy all the other queen cells, unless you want to use virgin queens in other hives. In this case remove the queen cells for future use.

If you miss a queen cell you will still run the risk of another cast. You may now have more than one virgin queen and they will sort themselves out leaving only one in the colony. A colony with more than one virgin queen will not swarm but miss one queen cell and a swarm or cast will probably result.

Collecting a Swarm

The old saying 'A swarm in May is worth a load of hay' is very true because it has time to develop into a productive colony during the summer. You should be able to get a brood chamber and super full of drawn comb and a honey crop by the end of July. Collecting and hiving a swarm can also be fun and provides a useful service for your local community.

More often than not you do not know the origin of a swarm. There is a risk that it may be carrying disease. Until recently I have been fortunate over many years and apart from some varroa infestation I have not yet collected a diseased swarm. Usually it is strong healthy colonies that swarm. Sadly I recently collected a swarm carrying EFB disease and it has cost me dearly. Always isolate a swarm until you know it is disease free!

You will need a skep, strong cardboard box or a nuc box and a sheet of smooth fabric large enough to wrap around the box. When you are off to collect the swarm you should take your veil and smoker. Ask the person reporting the swarm to describe its position and height. Be prepared to take secateurs, saw and loppers. Swarms land in all sorts of places. The ideal swarm lands at shoulder height on a single branch. Many swarms land in more awkward places.

Avoid taking silly risks yourself. Be safety conscious at all times towards helpers, onlookers and passers-by. Resist the temptation to climb a ladder to reach the one that is just out of reach from the ground. Explain to the owner of the property what you are going to do especially if you can see you will need to walk onto a flower bed or shake or cut off a small branch.

If you box a swarm during the day it is best to leave the box on the ground so that all the flying bees eventually enter it. Collect the box in late evening when all of the bees are inside.

Getting the Swarm in a Box

Spread the cotton sheet on the ground as near as you can to the swarm. Weight it down if necessary. If the swarm is hanging from a branch, hold the box under the cluster and give the branch a sharp shake to dislodge the bees into the box. If the swarm is on a thin branch cut off the branch and carefully deposit the bees in the box. Clustering bees will usually walk up into a dark space so if they are on a post or where you can position the box above the swarm, you can wait until they have entered the dark space inside the box.

When you have most of the cluster in the box place it upside down on the sheet and prop up one edge with a stone to create an entrance. By now there may be a lot of bees flying around and some will start to form another cluster on the branch. You can collect these in the same way with another small box and deposit them onto the sheet by the entrance. You may use the smoker to move them and give the site a good smoking to mask the smell of wax and bees remaining on the branch.

If you have the queen in the box you will see an increasing number of worker bees facing the entrance and raising their abdomens to expose their Nasonov glands and fanning. They are creating a flow of air to send the 'come and join us' pheromone into the surrounding area to guide the airborne bees to their new location. This is a good sign that the queen is present in your box. If the queen is not in the box there will be little or no fanning and you will soon see the bees fly away.

Foraging bees may still be returning during the rest of the day so the best time to collect the boxed swarm is after they have returned and stopped flying for the day if you take them home during the day the returning bees will form another smaller cluster the original site. Your neighbour will not be impressed, may be worried and will call you again. This time, without a queen it is not a swarm, and it will be impossible to collect the bees.

Hiving the Swarm

Before you go to collect the swarm in the evening prepare a hive with a full complement of frames and foundation. When you return to collect the swarm you will usually be greeted by a peaceful scene no bees to be seen and all inside. Carefully fold and secure the sheet over the box and take it to your prepared hive. It is a good idea to 'hive' the swarm as soon as you can because there is a risk of suffocation if the swarm is left in the box overnight.

Another quicker way is to dislodge the bees from the box straight into the hive after taking out the frames and temporarily closing the entrance. When you replace the frames don't press them down on top of the bees. Let the frames sink slowly under their own weight.

Prop up a board against the hive entrance and cover it with the sheet. Shake the bees out of the box on to it. Soon the bees will begin walking into the hive and if you watch you may see the queen. It may take up to an hour for all the bees to enter the hive.

Sometimes but not often the box will be empty! The bees, even with a queen, will have moved on. They are likely to have moved by up to a mile or two. Better luck next time!

Caring for a Swarm

The bees will get to work straight away making wax to build new comb. You should not feed the bees until they have used up the stores of honey in their honey crops. This will reduce the risk of disease carried in the honey being introduced to their new home.

After a 24 hours you should feed with sugar syrup. A good sized prime swarm will draw out a full box of foundation in a week or two. At this point check carefully for healthy larvae and sealed cells.

Build the colony up into as strong a one as possible. A swarm collected early in the season will build up strongly enough to be able to go to the heather moor. Alternatively build them up to go into the winter in peak condition ready for a great start to next year.

Bee Diseases & Pests

Honeybees are covered by the Veterinary Medicines Directive and the health of our bees is monitored by the National Bee Unit which is a part of the Animal and Plant Health Agency APHA - overseen by DEFRA based in York. The Bee Inspectors that work for the NBU are a great help to beekeepers and their service is free.

The National Bee Unit website BEEBASE is a must visit for all beekeepers having a wealth of advice and information freely available. It has details of all the pests and diseases that can affect your bees and most importantly it has a good picture library which shows you what to look for if you suspect there is something wrong with your bees.

Recognise the Healthy Colony

Recognising what a healthy colony and healthy comb looks and smells like is really important for the new beekeeper. If you know what is healthy then anything that does not seem to be healthy is something you should get checked out.

Visit BEEBASE and have a good look at the photographs of healthy combs of worker brood and larvae.

Note the essential features of healthy brood comb:-

Capped Cells (sealed brood – pupae) – the capping are dome shaped with a smooth(ish) surface and no pits or ragged holes.

Larvae are laying on their sides in the cells – a nice C-shaped, white and clearly segmented little 'maggot'.

New beekeepers are often confused by the appearance of pollen in the cells – different coloured polled can look suspect – check out pictures of pollen in combs too.

REMEMBER – If things look and smell healthy they probably are. If the colony is prospering and growing then things can't be too bad.

A Special Disease Inspection

If you want to make sure that you can spot diseases that affect the developing bees then you need to carry out a slightly different inspection of your bees. Use the same process as when you carry out a normal inspection but when you lift the brood box combs from the hive give the frame a sharp shake downwards so that most of the bees fall down into the hive. This will not harm the bees and will allow you to have good look at the comb and deep into the cells. Carefully replace the frame.

You might want to be extra careful with the frame that has the queen on. As you develop your confidence as a beekeeper you'll learn to be able to move the queen to another comb or to pop her into a cage for safety.

It is good practice to do three disease inspections a year; Spring summer and autumn.

The Diseases and Pests that cause problems

Bees have survived for 50 million years – Humans a mere 5 million years! If 50,000 humans lived together in close proximity in a small, enclosed, moist, dark, poorly ventilated environment, disease would be more prevalent. However...

- Prevention is better than cure.
- Learn to recognize healthy brood and bees.
- Adopt hygienic practices – do not discard comb & propolis in the apiary or exchange combs between colonies.
- Try not to squash bees when manipulating a colony.
- Avoid robbing and drifting – don't spill sugar syrup or manipulate late in the season.
- Assume second-hand equipment has contained a diseased colony. Blowtorch wooden parts or scrub polystyrene with washing soda and bleach solution. Do not use second-hand frames or combs.
- Quarantine swarms – check for disease.
- Never feed foreign honey or honey of unknown origin.
- Wash tool, smoker and gloves in 20% Washing Soda solution between hive inspections.
- If in doubt, ask for advice from an experienced beekeeper.
- Foul Brood diseases must be reported to the National Beekeeping Unit or the Bee Inspector.
- Diagnosis of Foul Brood and treatment is free.
- Don't always lament losses. If it is your mistake learn from the experience. Survival of the fittest
- and adaptation applies to bees.

Disease in this context is anything that causes the colony not to be at ease – diseased – not necessarily caused by a pathogen. Bee diseases may be grouped under Brood Diseases, Adult Diseases, Management and Pests:

Brood Diseases

It is essential to be able to recognise healthy brood – anything that deviates from this is suspect. A colony that is stressed is more likely to develop disease so please avoid stressing colonies. Avoid unhygienic procedures (e.g. exchanging combs, spilling sugar syrup) and don't over-inspect or manipulate your colonies. It is important that think about the needs of your bees combining what you know is happening inside the hive with what is happening outside the weather, forage availability and environmental conditions.

The main brood diseases are:-

Sacbrood (*Morator aetatulae*) is a virus disease found in 30% of colonies, usually noticed from May to early summer, when the ratio of brood to bees is high. Sacbrood is usually transitory and not a matter of concern. Combs can be re-used – the virus becomes non-infectious within a few weeks.

Chalk Brood is caused by the fungus *Ascophaera apis*. The dead larva is chalky white at first, often with a yellow centre, and becomes very hard and loose in the cell (mummies). Additional black/grey spores may develop on the surface. Mummies are removed by house bees and can be seen outside the hive or on the floor. Combs can be sterilized using acetic acid. Replace comb showing large amounts of chalk brood. Re-queening is recommended.

American Foul Brood is a disease of the PUPAE. It is caused by the spore forming bacterium Paenibacillus larvae. The larvae usually die after the cell is sealed. The comb has a pepper pot appearance where diseased larvae have been removed. Cappings may appear moist, sunken and perforated. Initially the dead larvae are slimy. It dries to form brown scales. The scales are difficult to remove and are highly infective – spores have been known to be viable after many years. A matchstick is inserted into a suspect cell, twisted and withdrawn slowly will pull out a brown mucus thread. AFB is a notifiable disease – the RBI confirm this using a lateral flow test. A standstill order will be put in place. If confirmed, the RBI will supervise the burning of bees and combs.

European Foul Brood is a disease of the LARVAE. It is caused by the bacterium *Melissococcus plutonius*. The bacteria feed on food in the larval gut and starve the larvae. Larvae usually die before the cell is sealed. Affected larvae are seen in unnatural positions ('stomach ache'), colour changes from pearly white to cream and eventually dry to form a brown scale removable by the bees. In early stages, infected larvae have a melted wax appearance. Cell contents do not rope. EFB is a notifiable disease. The Bee Inspector will confirm the diagnosis using a lateral flow device. Most colonies can be treated using the shook swarm technique.

Varroasis

Varroasis is not a disease but an infestation by the parasitic mite Varroa destructor. The mite is endemic throughout the U.K. and most of the world. Your colonies will have varroa mites.

Doing nothing is not an option – without treatment colonies will die within 3 years. You must learn to monitor colonies for levels of infestation and treat when necessary with the approved varroacides in the correct manner. Fit varroa screens under your mesh floors in order to monitor levels of infestation.

Treat varroasis using an integrated pest management system – consult the leaflet "Managing Varroa" obtainable from the National Bee Unit. This gives you very detailed advice and instructions about both monitoring and treating for varroa mites and should be a 'must read' for all beekeepers.
Local associations will also provide training and help for managing this pest.

Varroa mites breed in sealed brood cells – since a newly hived swarm has no brood, it can be treated to give a mite free start to the colony. Apart from seeing mites, you may see stunted bees with distorted wings, caused by deformed wing virus, resulting from the varroa mite sucking the larval haemolymph (blood) and spreading viruses – this is usually an indication of a high level of infestation. The puncturing of the larvae enables non-apparent viruses to take hold such as Slow Paralysis Virus and Deformed Wing Virus Acute, Chronic, Cloudy Wing Viruses – the colony dies from virus infection.

Visit the National Bee Unit website BEEBASE to get the latest and best advice on how to manage varroa infestations in your colonies. Learn to recognise the signs of varroa infection!!!!!! Learn how to do a varroa count!!!!!!

Varroa

Adult Diseases

Acarine is an infestation by the mite *Acarapis woodi*. Despite the signs of acarine given in beekeeping books, there are no visible external signs – the signs usually given, crawling bees, dislocated wings, etc. are those of Chronic Bee Paralysis. The mites infest the trachea. It usually has little effect in the active season. The mite is spread from old bees to very young bees. Treating your bees for varroa mites will also remove any acarine issues.

Nosema is a serious disease of the honeybees intestines, usually being seen in spring. The disease is caused by the spore forming microsporidians – *Nosema apis* and *Nosema ceranae.* Spores of these organisms can only be seen using a microscope. The obvious signs of nosema are faeces on combs and the outside of the hive.

Viruses. Nosema, acarine, varroa, etc. in themselves do not kill a colony they weaken it and thereby allow viral infections to take over. There are no cures for viral infection - antibiotics do not work on viruses. Viruses multiply in living cells of their hosts. In practice, most colonies terminally weakened with nosema or acarine exhibit signs of Chronic Bee Paralysis Virus, particularly clustering on top bars and continual trembling.

Management Issues

Chilled Brood is not caused by a pathogen. The optimum brood temperature is 35°C – 37°C. If there are insufficient bees to maintain this temperature, the brood will die. In the Spring the queen may have laid a patch of brood that the bees can't cover if the temperature drops. A characteristic is that brood of all stages, sealed and unsealed, are affected. The outer boundaries of the brood cluster are affected first as the bees retreat to maintain the inner core at the correct temperature.

Dysentery is not a disease but a condition caused by excessive build up of waste matter in the rectum i.e. diarrhoea. It is usually due to unripe honey/late feeding, granulated stores, fermenting stores, feeding brown sugar. The signs are fouling of combs, hive parts and around the entrance. Dysentery is often associated with nosema. A badly affected colony will be weakened and may succumb to viral infection. Soiled comb should be replaced and destroyed.

Poisoning. A sudden reduction in the number of foraging bees, a large number of dead or dying bees outside the hive, may indicate poisoning by bees alighting on sprayed crops. Legislation has reduced the number of incidents. Apart from the evidence of dead bees, the colony may become bad tempered, shivering, staggering and crawling bees may be seen. Rural beekeeping associations usually have a Spray Liaison Officer who will advise on the use of chemicals by local farmers. If you keep your bees near a farm it is worth chatting to the farmer who will usually let you know when sprays are going to be used. You can seal your hives for 24 hours to avoid major problems.

Pests that cause problems for Honeybees

The larvae of the Greater Wax Moth (*Galleria mellonella*) hatches among the brood combs and chews its way through cappings in straight lines. Stored comb is vulnerable to damage since the larvae feed on wax, larval skins and pollen. Protect stored comb by stacking boxes, placing newspaper between each box, and placing a saucer of acetic acid between every few boxes. Using acetic acid kills all stages. Greater Wax Moth has become more evident in recent years – maybe resulting from the loss of feral colonies and the use of varroa screens under which they pupate. Woodwork and polystyrene can both be damaged as the larvae scoop out boat-shaped depressions where they spin a cocoon and pupate.

The Lesser Moth (*Achroia grisella*) can cause similar problems.

Vandals / Cows / Horses. Vandalism of beehives and apiaries is uncommon. To cover yourself against prosecution by someone if they or their animals are stung by your bees you should put up notice letting people know that "Honeybees are Flying in the area Can Be Dangerous". Cows will mostly ignore bees but can push hives over. Bees and horses should never be mixed and a strong fence should keep them apart. Bees will collect horse sweat which annoys the animal causing it to kick the hive which in turns angers the bees.

Wasps and bumblebees. Autumn preparations should include taking precautions against pest damage to hives and colonies. Wasps and bumblebees should be kept out of hives by reducing the size of the hive entrance so that the bees can defend it more easily.

Mice /Woodpeckers / Rats / Badgers. In the winter all hives are at risk from mice and in many locations hives are also vulnerable to damage by woodpeckers and badgers, etc.

Fit mouse guards to the entrance as the first frosts approach to prevent mice getting into the hive. Larger pests can be deterred by wrapping the hive in chicken wire.

Beekeepers can be major pests of honeybees through their enthusiasm and careless mistakes. Opening hives in the cold, 'spreading brood', filling observation hives, moving bees, removing too much honey and so on.

If a colony dies the hive should be closed to prevent robbing and the cause of death ascertained. The combs from dead colonies need assessment. Dispose of old combs. Use a blowtorch to sterilise woodwork. Colonies that appear sick, e.g. not building-up in the spring, should be left alone. Feeding may help if they are short of food. Otherwise, give them a small entrance and leave them alone.

Foraging

Honeybees collect nectar from flowers. They carry this in the 'honey stomach'. Nectar is turned into honey by the action of enzymes secreted into by the bee and by evaporating most of the water from it.

Bees collect pollen as a source of protein to feed to the young. They also store pollen. Pollen is carried (as 'bee bread' mixed with nectar) in the pollen basket.

Honeybees are attracted to many plants which provide them with nectar and pollen in exchange for the their services as pollinators. The main nectar sources are:

Agricultural
Oilseed Rape	Linseed	Field Beans	Sunflowers	Fruit crops
Borage				

Hedge rows
Hawthorn	Bramble	Wild Rose	Poppies	Dandelion
Water Balsam	Ivy	Rosebay Willow Herb		

Trees
Lime	Chestnut	Sycamore	Fruit	Horse Chestnut

Parks & Gardens
Most trees and native plants with single flowers

Heather Moors
Ling Heather

Flying Distance

Bees fly up to 3 miles in any direction to collect nectar and pollen. They fly directly from hive to plant and back. Have a good look at the area in a three mile radius of where you keep your bees using any of the mapping software available. This will give you a good idea of the forage available to your bees. Rural areas often lack a range of forage sources while suburban areas are often abundant in forage sources. Consider what forage is available from April to September.

Nectar

A single bee makes thousands of journeys to collect Nectar. Nectar contains between 40 - 80% water while honey contains 15 - 21%. One kg of honey takes approximately 100,000 bee journeys. Nectar provides the carbohydrates need by your bees and so is the energy fuel for everything they do. Sugars are stored as honey but are also converted into wax for comb building.

Pollen

A strong colony requires approximately 10Kg of pollen to raise 100,000 bees - a million loads of pollen from around a variety of plants. Pollen provides most of the rest of the bees dietary needs especially the protein needed for brood rearing. Pollen also supplies minerals, vitamins and roughage. A bees diet needs to contain a variety of pollens to ensure they get all of the essential amino acids they need.

Water

Honeybees need a source of water. This is essential to their well being.

Honey

A strong colony might produce 200 Kg of honey in a good season. They need 160 Kg for their own use to build comb, keep the colony warm and to feed both young and adults They need at least another 22 Kg to provide them with enough stores to survive the winter. The beekeeper can harvest any honey made in addition to this. Much of the art of beekeeping is in persuading the bees to produce much more honey than they would in nature. As partners the human provides the house and keeps the bees healthy while the bees provide the honey.

Pollination

Honeybees, along with many other insects and other animals are really important pollinators. Almost everything we eat other than cereals requires pollination by insects. Honeybees are particularly important in the agricultural setting where a large number of pollinators are needed. Having honeybees in the field of oil seed rape, in a good year, will increase its yield by 15%. Fruit crops improve both in quality and quantity if honeybee colonies are present.

Dancing Bees

Bees communicate to one another the position of good food sources by performing a range of dances. The 'round dance' indicates food sources close to the hive. The waggle dance gives exact instructions of where foragers should fly to to find prime food sources. You don't need to know about this to keep bees but you really should find out a lot more about it because it is amazingly clever of your bees.

BEE PLANTS AND FLOWERS

Honeybees will collect nectar and pollen from single, simple and short flowers. The flowers will only produce the nectar if environmental conditions are suitable. This list gives an idea of which plants are useful for honey production and colony health. Take the details with a pinch of salt because climate change has really altered flowering times. Just keep you fingers crossed that conditions for flowering and your bees flying actually coincide.

	From	To	Pollen Yield	Nectar Yield
Winter Aconite	January	April	M	P
Crocus	February	April	G	P
Dandelion	February	October	G	M
Dead nettle	February	September	G	P
Hazel	February	March	G	-
Hellebore,	February	May	G	P
Hyacinth	February	April	M	M
Snowdrop	February	March	M	M
Blackthorn	March	April	M	M
Gorse	March	June	G	P
Grape Hyacinth	March	May	M	M
Willow	March	May	G	M
Almond	April	May	M	G
Anemone	April	June	M	M
Bluebell	April	June	M	M
Gooseberry	April	May	P	M
Horse Chestnut	April	June	G	G
Lilac	April	May	P	P
Apple	May	June	G	G
Berberis	May	June	M	M
Basil	May	July	M	M
Beech	May	June	M	-

Notes for new beekeepers

Plant	Start	End	Nectar	Pollen
Broom	May	June	G	M
Cherry	May	May	G	M
Clover	May	July	M	G
Cotoneaster	May	August	P	M
Forget-me-not	May	July	M	G
Holly	May	July	M	G
Hawthorn	May	June	M	G
Sycamore	May	May	M	G
Mountain Ash	May	June	P	P
Mustard	May	June	G	G
Pear	May	May	G	G
Raspberry	May	June	M	G
Strawberry	May	June	G	M
Field Beans	May	June	M	G
Oil Seed Rape	May	June	G	G
Blackberry	June	August	M	G
Borage	June	November	M	G
Brassica	June	July	G	G
Lavender	June	July	M	M
Lily of the Valley	June	August	M	-
Common Lime	July	August	M	G
Marjoram	July	August	M	G
Meadowsweet	July	August	G	M
Saxifrage	July	August	M	M
Rosebay Willow herb	July	September	G	G
Himalayan Balsam	August	September	G	G
Heather	August	September	G	G
Ivy	September	December	M	P

Key

G = Good   M = Moderate   P = Poor

Feeding Honeybees

There are 3 situations when we might need to feed our bees a sugar supplement:-

Spring or Stimulus Feeding

Spring is the time a colony might starve to death, especially if the winter is long, cold and wet. Feeding fondant in the cold is the best source of food for bees BUT when spring finally arrives you can give your colonies a boost by giving a stimulus feed. This consists of a 1:1 sugar water mix. Adding some pollen or pollen supplement will also help.

In an ideal world the colonies will be situated near a good water source surrounded by plants like willow and hazel that give early pollen together with early nectar producing flowers. This will give bees the best start to the summer. However if you want your colonies ready for a big nectar flow such as oil seed rape you can use stimulus feeding to get colony numbers up very quickly so that you can get the biggest possible harvest.

Emergency Feeding

In our changeable weather bees can be in danger of starvation at any time of year. Big colonies are in the greatest danger. If the colony is running out of food stores in winter then give a supplement of white fondant candy. If there is suddenly 3 weeks of rain in summer then feed your bees with a strong 1:2 water:white sugar solution.

Autumn Feeding

A small colony will need a minimum of 10Kg of stored honey to survive the winter while a strong colony might require more than 20Kg. Most beekeepers remove a honey crop in August giving the bees time to build up stores in the brood box but with poor weather the bees are often not able to fill their stores with enough food for winter. In this case the beekeeper must supplement the feed by giving the bees a feed of strong 1:2 sugar solution. You should aim to feed at least 10 litres of syrup to each colony.

As a guide each brood frame full of stores will weight around 2.2Kg so your brood box need 10 frames full of stores for the bees over the winter period.

When to Feed

Even if you have just one colony it is best to put feed on the hive in the evening. This prevents other bees and insects from robbing the colony and possibly killing it. Ideally put an entrance block in place so that the colony has a smaller entrance to defend.

What to Feed

Only feed your bees with pure white granulated sugar or fondant candy. You can add pollen or pollen supplement to the feed but do not add honey or any form of brown sugar.

Fondant Candy

You can buy bakers fondant and fondant especially formulated for bees or you can make your own. This is used as a winter feed. Cut a hole in the bag it comes in or fill a container with candy. Place this either (1) under the roof directly over the hole in the crown board so that the bees can climb up to reach it or (2) remove the crown board and place the fondant on top of the frames directly above the cluster of bees. In both cases you may need an eke, (1) above the crown board or (2) below the crown board, to make room for the fondant.

Sugar Syrup

You can buy sugar syrup designed for bees or you can make your own, which is cheaper but messier in your kitchen. If making your own dissolve 1Kg of white sugar in 630ml of water. You'll need to heat it to almost boiling point but don't caramelise it. Store in a clean plastic container.

How to Feed

Rapid Feeders

A tray of some kind which sits at the top of the hive so that the bees can climb up to reach the food. These can be filled without releasing any bees from the top of the hive. Miller and Ashfield feeders replace the crown board but most beekeepers now use a plastic tray that sits over the central hole in the crown board under the roof. As a general rule the bigger the feeder the better.

You may need an eke to allow the feeder to fit under the roof. Best used for autumn feeding.

Contact Feeders

Basically a bucket with a gauze grid in the centre of the lid. Fill the bucket with feed and quickly turn it upside down. A partial vacuum is created stopping the food leaking out. Place this over the hole in the crown board. You'll need a empty super or brood box to act as an eke below the roof.

Best used for spring and emergency feeding.

Frame feeders

A feeder the size and shape of a frame. It hangs in the hive just like any other frame. A good way of getting some feed right next to the bees. Used in small colonies. You must use a float in order to prevent bees drowning.

Hive Products - Honey, Wax, Propolis and Pollen

Honey
Honey is stored in the comb by the worker bees. Its storage is a model of hygienic food preservation, each cell is filled with well ripened honey and covered with an individual wax capping.

Most honey in the jar is a blend containing a mixture of nectars gathered in the area round the hive. Honey varies in colour. Willow herb is one of the lightest, the range goes through the golden shades from a pale straw to strong hues, darker colours come from hawthorn and field beans. When taken from the hive fully sealed most honey is liquid but after a period of storage, particularly after extraction, it will granulate.

Granulation is a normal feature of honey. Clover honey granulates with a fine smooth quality. Some types of honey set rather hard with a coarse structure, which can be avoided by a process of warming and stirring-in some partly melted crystallised honey of a better type as a seed to give a smooth texture. Later on the honey can be warmed and stirred to produce creamed or soft set. Both soft set and granulated can be made clear again by warming gently.

Heather honey is in a separate class, it is a jelly with bubbles in it which liquefies when stirred, it is usually a rich reddish amber in colour.

Comb honey can be produced using starter strips of wax in your super frames or by using specific cut comb foundation. Once the bees have built comb, filled it with honey and sealed it in you can cut out the comb to either pack in cut comb containers or to put in jars of liquid honey as 'Chunk Honey". Alternatively you can produce small wooden frames filled with comb called sections. These are fairly difficult to produce. Cut comb honey retains its full flavour and aroma of the natural product. This is the ultimate 'raw' honey.

Wax
Beeswax is a premium product that can be used to make polish, candles, cosmetics as well as many other products. If you do not want to use the cleaned wax yourself you can exchange it for fresh foundation.

Propolis
Propolis (bee glue) can be collected from the hive. It is used to make a tincture that can be found in many health food shops.

Pollen
Some beekeepers collect pollen which is used in hay fever treatments.

Royal Jelly
Again this can be collected and sold to pharmaceutical companies. It is used in beauty and ageing treatments.

98 Notes for new beekeepers

Pollen in frame

Honey

When does nectar become honey ?
Plants use water, nutrients from the soil and carbon dioxide from the air in order to manufacture their food. They do this by photosynthesis. Pigments including green chlorophyll present in the cells use light energy to split some of the water to produce hydrogen ions which can be joined to carbon dioxide from the atmosphere to form sugars. These can then be converted in many more complex chemicals. The nectar secreted by the flowers contains sugars in varying proportions according to the species of the plant, the soil and climatic conditions. The nectar also contains traces of protein, salts, acids, enzymes and aromatic substances all in a watery solution. Nectars vary considerably in flavour and sweetness.

Nectar is secreted by a flower at specific times of the day according to the species of the plant. Its secretion is influenced by temperature and humidity as well as by soil moisture content. Honeybees collect nectar as a reward for providing pollination services for the plant. It is carried back to the hive in the bee's honey sac, a non digestive crop. The foraging bees pass the honey to younger bees who will place it in a comb or use it for food.

Two things have to be done to convert the nectar into honey. The water content of around 80% has to be reduced to around 18% to prevent fermentation. Bees do this by warming the nectar to 33°C. The sucrose in the nectar is converted by the enzyme invertase into two sugars fructose and glucose. The concentrated honey is then sealed in the cells of the comb and will keep until needed. Bees add water to the honey before feeding it to larvae.

Oil Seed Rape Honey
Oil Seed Rape yields very large quantities of honey but this honey granulates into a solid 'concrete' honey if left to itself. The crop can be removed prior to capping, extracted and have some of the water content removed before creaming and jarring. Rape honey left to set in the supers can be melted to yield liquid honey and wax.

Borage Honey
Borage also gives very high honey yields but care must be taken to ensure that only fully capped honey is extracted. This watery honey will ferment easily if the honey is not fully ripe. Borage honey does not granulate quickly.

Heather Honey
Heather honey is too viscous to extract using a centrifugal extractor. It must either be extracted in a heather press or it can be sold as cut comb honey.

A word about Comb Honey
Thin unwired foundation or starter strips are used to encourage the bees to produce new comb. When filled with honey and capped this is removed from the hive and cut into sections for sale as cut comb. You eat the whole honeycomb, wax and honey.

Extracting and Bottling Techniques

Clearing Honey Supers of Bees
Honey supers in the apiary are full of bees. To remove the bees from the super so that you can take the frames to the extraction room you need to place a clearing board with a bee escape below the honey box. At night the bees will move down through the escape into the brood chamber. The escape will prevent them returning to the honey box the next morning. Whole books have been written about 'clearing' and lots of different techniques, boards and escapes exist for you to try. Each beekeeper has their own preferred method. I love my rhombus escapes.

Uncapping
Removing the wax seal from across the top of the honey filled honeycomb. Use a knife or an uncapping fork. Hold frames over a tray/dish. A warm frame will be uncapped more easily. Many expensive devices are available to make this easier and less sticky.

Centrifugal Extraction
Most honeys can be extracted by spinning the frames of honeycomb in an extractor.
Set honey and heather honey cannot be extracted in a spinning extractor.

Straining
The extracted honey will contain bits of bee and a fair amount of wax. Remove these by straining through a fine sieve and some nylon net (double strainer and straining cloth)

Settling
When honey is extracted it gets lots of air bubbles in it. Allow the honey to settle for 24 hours before bottling it.

Ripening

Honey that is a little too 'watery' can be improved by gently heating it to evaporate water from it.

Refractometry

You can test the water content of honey using a refractometer. Honey should have a water content between 17% and 21%.

Selling Honey

Honey is a high quality product and should be prepared for sale with care so that customers get the best quality product available. Selling to family and friends is fairly free of legal requirements but as soon as you start to sell in shops, markets or via the internet you will need to conform to food hygiene regulations and labelling laws. Have a look at Hazard Assessment and Critical Control Points documentation to find out about the key points for producing honey in a hygenic manner according to food regulations.

Source and Clarity
Honey must be exactly what is stated on the label e.g. Pure Yorkshire Honey.
It should be 'clean and clear'.

Containers
The size and quality of containers, and the weights of honey that can be sold are defined by law.

Price
Ask other beekeepers what they charge and set your price at the same sort of level.

Labelling Law
The law clearly states what should be on a label and how big each element should be.
You must use tamper evident labels if selling honey to the public.

Marketing
You will have no trouble selling your honey to friends and to the public via shops and markets.

Association Sales
Many beekeeping associations will invite you to sell your honey at local association shows and events.

THE BEEKEEPING YEAR

Late August
Start to prepare the colonies for winter. For a colony to over winter successfully make sure that:
1. The colony is disease free - COMMENCE VARROA TREATMENT NOW
2. The colony has a young fertile queen
3. The bees have 22 Kg of honey stores
4. The hive is weatherproof and on a stable stand.
5. The bees are protected from pests such as mice.

After the honey supers have been removed examine the colony and determine the quantity of stores in the brood chamber. Feed those colonies which do not have sufficient stores. As a rough guide one brood comb will contain around 5 lbs of sealed honey when full.

Add an entrance block to reduce the size of the entrance. Robbing can be a serious problem with strong colonies robbing weaker ones. Wasps will rob hives and can destroy colonies so put out wasp traps if there is a big wasp population in the area.

Weak colonies can be united and colonies can be requeened.

September / October
Colonies should be collected from the heather moors. They will need to be checked for stores and given a varroa treatment.

Make sure
1. You have fitted mouse guards
2. The hive has adequate ventilation.
3. You have secured your hive so that it will not be blown or knocked over.

November / December - January

Apiaries should be inspected fortnightly, or after particularly bad weather, to ensure that all is well. The hives should not be touched, even taking the roof off will lower the temperature of the cluster unnecessarily. Heavy snow can block hive entrances so clear this away.

In late December or early January treat your colony with oxalic acid to remove varroa mites. The colony is broodless so all of the mites will be on the adult bees. Dribbling or fumigating the colony with oxalic acid will remove more than 90% of the mites. Follow instructions carefully as oxalic acid is toxic to humans.

Late March / Mid April

What needs to be done depends on the weather. On a warm day an examination can be carried out as follows.

1. Mouse guards should be removed.
2. Start hive record for that year.
3. Check food Stores and feed if necessary.
4. As the weather improves they will require a nearby supply of water to dilute stored honey.

April in May depending on the weather

Nectar becomes more available during this month. If the weather is warm begin regular colony inspections. Colonies are small at this time of year so it is the ideal time and therefore the best time to do those spring cleaning' tasks:-

1. Mark or remark queens and clip wings
2. Check the hive for any repairs that are needed
3. Old comb at the outer edges of the brood box should be replaced.
4. Change or clean the floor.
5. Supers should be added as soon as the two outside frames are the only ones without bees.
6. Any colony which is slow to build up should be examined for a possible reason.

May

Usually a very busy beekeeping month.

Weekly inspections for colony management and swarm control

1. Are there sufficient stores to last until the next inspection?
2. Is the queen present and laying normally?
3. Are there any signs of disease?
4. Is there sufficient space in which the queen can lay?
5. Are the bees preparing to swarm?
6. Does the colony need additional honey supers?

When you become more experienced this month is a good time to start queen rearing and producing nucleus colonies.

If the weather is good and you live near plentiful spring nectar forage then you can consider extracting a crop of spring honey - but remember that even in summer the bees need enough honey to survive a period of poor weather.

June

The colony needs to be as large as possible so that they are able to collect as much forage as possible. The colony should still be expanding and further supers may be necessary. Traditionally this month has a reputation of having a June gap due to a lack of flowers producing nectar but with climate change this is no longer true. The nectar flow is very changeable and you will need to match your beekeeping to the weather and the availability of flowering plants. If the weather is poor colonies may be short of stores and you may have to feed them.

July

July should be a good month for forage with lots of plants in bloom. You should continue with swarm control inspections and adding supers if needed. This is also a good time to move frames around in supers as bees tend to fill the middle frames and ignore the outer ones - swap them over so the bees are filling all the frames.

With three or four supers on the colonies it is a hard job lifting them off for swarm control.

If you plan to take bees to the heather moor you should start to prepare them now. The essentials for the heather stock are:

1. A young queen that is laying really well
2. A very full brood chamber with brood on all frames.
3. Enough stores to see it through until the heather flowers
4. Drawn comb is useful in the supers, as it is usually colder on the moors

August

When the amount of flower available on the local flora drops to a low level then you should extract any capped honey.

Wet supers (supers filled with frames that you've just extracted the honey from) should be returned to the hives they came from. Place them above the crown board leaving the opening so that the bees can collect the honey from the wet supers and store it in the main hive. Some beekeepers prefer to store honey supers wet as this deters wax moth.

If you live in an area blessed/cursed with large areas of Himalayan balsam then your bees may well go on storing honey well into September. It is important to relate your beekeeping activities to the weather, the local flora and most importantly the needs of your bees.

	Jan 1	Jan 15	Feb 1	Feb 15	Mar 1	Mar 15	Apr 1	Apr 15	May 1	May 15	Jun 1	Jun 15	Jul 1	Jul 15	Aug 1	Aug 15	Sept 1	Sept 15	After Door Oct 1	Oct 15	Nov 1	Nov 15	Dec 1	Dec 15
Management	Management to maintain colony integrity								Active Management to increase colony size and honey production										Management to maintain colony integrity					
Feeding		Candy if Needed						Stimulus									Feed colonies ready for winter							
Supering									Add supers in advance of brood development - nectar takes up lots of space															
Requeening Uniting											Requeening						Unite Weak Colonies							
Swarm Control									Look for queen cells - have equipment ready for artificial swarms															
Mark & Clip									While colonies are small															
Bee Breeding							Drone Rearing		Grafting and queen mating						Build up nucs for overwintering									
Forage							OSR Willow Hazel Flowering Current	OSR Willow Hazel Flowering Current	Dandilion Fruit trees Field Bean Cotoneaster	Hawthorn Sycamore Oak Horse Ch.			Bramble Willowherb	Bramble Willowherb	Balsam	Balsam Heather	Balsam Heather	Balsam Heather	Ivy					
OSR Prep						Pollen Supl.	Bees on Rape followed by Field Beans																	
Heather Prep													Build Colonies ready for Heather				Bees on Heather Moor							
Varroa Control	Oxalic Acid							Thymol					Dusting with Icing Sugar at anytime				Thymol	Thymol						
								MAQS	MAQS	MAQS	MAQS	MAQS	MAQS	MAQS	MAQS	MAQS	MAQS							
Nosemma Testing																	Test for Nosemma							
Number of Adult Bees in Colony	20000	15000	16000	10000	12000	15000	20000	25000	30000	40000	50000	60000	70000	80000	80000	70000	60000	50000	45000	40000	35000	30000	28000	25000
Brood Numbers	0	100	200	500	3000	7000	10000	15000	20000	25000	25000	30000	30000	30000	20000	15000	10000	2000	0	0	0	0	0	0
Extracting								Extract OSR							Extract Main Crop		Heather Honey							

Summary for a year

New Beekeepers Checklist

This checklist is designed to give new beekeepers a list of the basic skills they will need to become a reasonably good novice beekeeper. It also lists the knowledge you will need in order to keep your bees to a good standard.

Consider these things before you get your bees and during your first year as a new beekeeper.

Have I attended a course that gives me both theoretical knowledge and hands on training?
Have I joined a local beekeeping association?
Am I aware of the time commitment needed to care for bees throughout the summer?

Is my apiary in a good forage area?
Is my apiary set up with the safety and health of the bees in mind?
Is my apiary set up with the safety of my family and the public (and animals) in mind?
Is my apiary set up to make life easy and safe for me as a beekeeper?
Are all parts of my hive equipment correctly built and setup to ensure bee spaces are maintained correctly?

Do I have the details ready to hand if I ever need to call the emergency services?

Is my beesuit in good repair and washed regularly?
Can I light a smoker and keep it alight during a hive inspection?
Can I use a smoker to calm bees rather than to make them aggressive?
Can I use a hive tool correctly?

Do I know the names and functions of all the hive parts?
Can I build hive parts and frames correctly?
Are my hives well maintained and arranged well?

Am I able to:-
inspect a colony while keeping everything hygienic?
lift a frame from the hive without 'rolling' or squashing bees?
identify Workers, Drones and Queens?
identify stored honey and pollen?
identify eggs, larvae and pupae (sealed brood)?
keep accurate hive records and use them to plan future inspections and work that needs doing?
recognise queen cups, open queen cells and sealed queen cells?
have equipment ready in case my bees decide they wish to reproduce by swarming?
carry out a basic of method of swarm control?
ask for help if I have problems?

Do I:-
know what healthy brood looks like?
know how to carry out a disease inspection of brood combs?
know who to seek help from if I suspect my bees have disease?
have a plan for controlling the levels of varroa mite in my colonies?

If I get a honey crop do I know how to remove and extract it?
Am I able to ask for help of I need to learn techniques?

You might also want to look at the syllabus for the BBKA Basic Examination.
You can take this assessment after keeping bees for a year.

A Final Warning

Beekeeping is addictive. Once you become fascinated by the thousands of bees in their hive you'll find it hard do stop being a beekeeper.

Photographs by members of Bradford Beekeepers Association

Drawings by Paul Hudson

Hive Record Card

Hive No:

Date

Inspection carried out by:-

Brood Box Record

Frame	1	2	3	4	5	6	7	8	9	10	11
Queen											
Queen Cell											
Eggs											
Larvae											
Pupae (sealed brood)											
Pollen											
Nectar											
Honey											
Frame with Foundation											
Drawn Comb											
Empty Drawn Comb											
Varroa/Disease ?											

Configuration (at end of inspection)

Honey Supers?

Observations & Actions Today

Preparation for next Inspection